A REVIEW OF THE BOOK

"A useful book for all signal engineers, can be a reference book for all courses in Indian Railway training institutes, will benefit those engineers engaged in research, design, setting standards, safe operation and maintenance of electronic signalling systems"
— Somnath Pal

ELECTRONIC INTERLOCKING
ON INDIAN RAILWAYS
WITH A SAFETY-CRITICAL ANALYSIS

PURNACHANDRA RAO VALLABHANENI

INDIA • SINGAPORE • MALAYSIA

Notion Press

Old No. 38, New No. 6
McNichols Road, Chetpet
Chennai - 600 031

First Published by Notion Press 2018
Copyright © Purnachandra Rao Vallabhaneni 2018
All Rights Reserved.

ISBN 978-1-64324-271-2

CONTENTS

PREFACE

It is the intention of this book to bring forth both the practical and theoretical aspects of 'Electronic Interlocking' including the requirements of safety and reliability. As the vital outputs of the electronic system depend on the correct operation of the software, more emphasis has to be laid on the validation of software. The constraints set for safety in just 'recommended' programming languages like 'C' have to be rigorously implemented through as many levels of checks as possible. The principles of theoretical computer science have to be applied to verify the railway interlocking and to bring out hidden errors which cannot be traced by applying traditional methods of simulation and testing. These aspects have been covered in the Chapters 3 to 9.

The Chapter 1 traces the brief history of the Indian Railways from the time the first train ran in 1853 till now. Four gauges of railways existed till the 1990s and now the only gauge with a few exceptions is the broad gauge (5'6"). The gauge chosen for urban transport is the standard gauge (4' 8½"). In the year 1947 there were 42 railways managed independently, which were consolidated into 6 zonal railways initially after nationalization but later bifurcated into 16 railway zones for better management and to meet regional aspirations. For managing specialized activities there are about 13 corporations under the Ministry of Railways. The evolution of signalling from the provision of just two signals (main and outer) without interlocking to advanced stages of provision of route relay interlocking and electronic interlocking has been described in Chapter 2.

The details of the Electronic Interlocking Systems (4 nos.) approved for installation on Indian Railways are portrayed in Chapter 3. The other systems (3 nos.) installed earlier but not approved at present are detailed in Appendix B. In Chapter 4, the aspects of safety and reliability have been discussed in detail. It is observed that a single processor system with error detection and diagnostic routines is economical as the enhancement in reliability in double and triple modular systems is marginal. Rigorous error detection methods to enhance safety are suggested. It is emphasized that qualitative methods to improve safety are required instead of relying only on quantitative figures for accident probability or hazard rate. Quality and safety management in the Indian context has been detailed.

Some methods of hazard analysis – Failure, Mode, Effects and Criticality Analysis (FMECA), Event Tree Analysis (ETA), Fault Tree Analysis (FTA) and Markov Analysis have been applied on some physical systems and the results analyzed in Chapter 5. Application of Binary Decision Diagrams (BDDs) for verification of hardware circuits and Verilog Hardware Description Language

(Verilog HDL) for verification of microprocessor circuits have been explained in Chapter 6. Non-functional hardware testing and application of Higher Order Logic (HOL) have also been described.

Verification and Validation (V & V) of software including testing has been described in Chapter 7. Safe subset of 'C' language and simulation of signalling at a yard in 'C' language have been detailed. Verification of 'C' programs with CBMC (C Bounded Model Checker) has been described with examples. In Chapter 8, verification of railway interlocking through Symbolic Model Verifier (SMV) has been dealt and a control table has been generated with the application of Ruby language and SMV. Application of Systems Engineering to Electronic Signalling Systems in India has been described in Chapter 9 giving importance to the safety aspects.

Acknowledgements

I am grateful to Sri P.V. Muralikrishna Sr. Professor Indian Railways Institute of Signal Engineering and Telecom. Ministry of Railways at Secunderabad and his deputy Sri Koteswara Rao for partaking of the latest information about the Electronic Interlocking (EI) Systems installed on Indian Railways. I thank Smt. Kalavathi Chief Signal & Telecom Engineer (Planning) Southern Railway for releasing some signalling plans to help our studies. I am also grateful to Sri Alok Katiyar Director/Signalling III of Research, Design and Standards Organization (RDSO) of Ministry of Railways for giving some non-proprietary information about Electronic Interlocking. I am indebted to Dr. Vijaya Kumar, Chief Signal and Telecom Engineer (Projects)/S.Rly for permitting the project team of SSN College of Engg. to visit a newly installed EI system.

I would like to express my gratitude to Mr. K. Sriram graduate student and Dr. S. Sheerazuddin Assoc. Professor Comp. Science of SSN College of Engg. for undertaking the work of verification of signalling and interlocking at a typical station and assisting me in the application of the theory of Symbolic Model Verification (SMV). Also I am grateful to Mr R. Keshav an undergraduate student of SSN College of Engg. for helping me in applying CBMC ('C' Bounded Model Checker) Tool developed by Carnegie Mellon University USA.

FEEDBACK

However carefully the book is written, errors cannot be completely avoided and there is always scope for improvement. I would therefore welcome any feedback and I can be contacted by e-mail at purna_chaandra@aol.in.

Pumachandra Rao Vallabhaneni

Chennai, India Oct. 2017

CHAPTER **1**

BRIEF HISTORY OF INDIAN RAILWAYS

1.1 Introduction

Before going into the details of electronic interlocking systems on the Indian Railways including safety etc. it is informative to know about the introduction of railways in India and developments in signalling over a period of more than 160 years brought out in the first two chapters of this book.

1.2 Introduction of a new form of transport in India

The need for a railway in India was felt by the British who controlled India through their East India Company which was originally a trading company, for transporting cotton to the ports for export to Britain and for better transport of their troops over large parts of the subcontinent. On Aug. 1 1849, the Great India Peninsula Railway Company (GIPR) was incorporated and was contracted to construct an experimental line between Bombay and Callian (Kalyan) and beyond. First official or 'ceremonial' train was run on Apr 16 1853 from Bombay to Thane (34 km) and this date remains as the date of introduction of railways in India. Two other companies – the East Indian Railway and the Madras Railway were contracted in 1850 for development of railways around Calcutta and Madras, respectively. The first train on the eastern side was run between Howrah and Hoogly (24 kms) on Aug. 15 1854 and in the south, between Royapuram and Wallajah Nagar (63 miles) on June 26 1856. These companies and those incorporated later mostly in England were guaranteed a fixed rate of interest on the capital invested, by the then East India company, for them to survive the climatic and commercial conditions then met in India.

1.2.1 Multiple gauges

A gauge of 5 feet 6 inches for the railway track all over India was selected by the then Governor General Lord Dalhousie, but later, lesser gauges i.e. the metre gauge (1 metre), narrow gauges (2 feet 6 inches and 2 feet) were approved by Lord Mayo then Viceroy around 1862 for installation in light traffic areas with connections to the broad gauge (5'6"). This multigauge network prevailed till wholesale conversions took place in the 1990s based on 'uniguage' policy. The route kms as on

31.3.2014 are B.G.(Broad Gauge) – 58175 kms, M.G. (Meter Gauge) – 5334 kms, NG (Narrow Gauge) – 229 kms, in comparison to the values in 1951, namely BG-25258 kms, MG – 24185 kms, NG – 4153 kms. Gauge conversion from M.G. to B.G. is still in progress to eliminate the MG completely except for some mountain railways.

1.2.2 Examples of Narrow Gauge lines

Examples of early NG railways were the Gaikwad Baroda State Railway (GBSR) of 2 feet 6 inches gauge and the Darjeeling Himalayan Railway (DHR) of 2 feet gauge. The GBSR was hastily completed in 1863 by the Princely state of Baroda to transport cotton to the markets in England by using bullocks to pull the wagons till the arrival of steam locomotives from England in 1873. This was the first NG line in India. The DHR from Siliguri to Darjeeling (a hill station) of 51 miles in length was opened in July 1881. This line is considered an engineering feat with a number of steep gradients and amazing loops and was designated as a World Heritage Site on Dec. 2 1999 by UNESCO.

1.3 Expansion of the Railways

The network of railways expanded towards the North also with the first line from Allahabad to Kanpur (119 miles) commissioned on March 3 1859. It continued to expand yearly from the 1860s to the first decade of the 20th century. By 1880, the railway network had a route mileage of 9000 (14, 500 km), mostly radiating inward from the three major port cities of Bombay, Madras and Calcutta. Railways were developed not only in the British administered parts of India but also in the regions of the various independent kingdoms, for example, the Nizam State Railway and the Mysore State Railway. There were thirty two Princely state railways, many of them of low route mileage, the shortest being the Dholpur State Railway with only 56 miles. In 1947, there were 42 separate railways including the 32 of Princely state railways with overall 55000 route kms of multigauge excluding the railways coming under Pakistan (1947 was the year of independence for both the countries).

1.3.1 Management of the vast network of railways

To coordinate the working of the many railway systems and formulation of time tables the Railway Board was formed in 1905. In 1925, the railway finances were separated from the general finances and a separate railway budget was presented, which practice continued till 2017. To coordinate the charging, booking and carriage of passengers and goods over the various systems, the Indian Railway Conference Association (IRCA) was formed in 1902 and made permanent in 1926 when it was also tasked to coordinate wagon maintenance and classification. The World wars, the first (1914–18) and the second (1939–45) had devastating effects on the Indian Railways as some of the assets such as locomotives, wagons and track material and the output of railway workshops were taken out of India for the military needs of the allied forces. In 1946, all the remaining company railways were taken over by the Government to speed up rehabilitation. GIPR, however, was taken over in 1900 itself.

1.3.2 Locomotives, coaches etc

A steam locomotive for the MG was the first to be manufactured in 1895 at the Railway Workshops at Ajmer, part of Rajputana – Malwa Railway. However, most of the locomotives were imported from England till 1947. The first of the indigenous steam locomotives (named Deshabandhu) of the WG class was commissioned on Nov.1 1950 from the Chittaranjan Locomotive Works. The railways' requirement of locomotives came out from these works till 1986, when the steam traction was phased out. From 2000 onwards, only diesel and electric locomotives are used for hauling the passengers and goods. These locomotives are also indigenously produced at two designated factories at Varanasi and Chittaranjan. The passenger coaches required are being manufactured at two factories at Madras (Chennai) and Kapurthala. The goods wagons have been modernised and are being manufactured in industries mostly located in eastern India.

1.4 Electrification of railways

The first electric train ran between Bombay and Kurla (16 kms) on Feb.3 1925 powered by 1500V D.C overhead traction and similarly some suburban sections around metro cities were electrified between 1925 and 1930. In 1956, the Government decided to adopt 25 kV A.C. single phase traction as a standard all over India to meet the challenge of growing traffic. The first section with this traction was Burdwan – Mughalsarai and subsequently most of the main lines are electrified, with total electrified Route kilometrage on BG exceeding 25, 000.

1.5 Consolidation of the Railways post 1947

The country gained its independence in 1947 and was declared a constitutional republic in 1950. The 42 railways including 32 princely state railways were consolidated and divided into just six railway zones in the years 1951 and 1952. These were further subdivided into nine zones between 1955 and 1966. These nine zones were further expanded to 16 zones in 2002 and 2003, which figure stands till now. The route kilometres as on 31.3.2014 is 65, 806. A list of the present railway zones is given in Table 1. A map of Indian Railways can be seen at www.indianrailways.gov.in

1.6 Development of Urban Transport

To solve the problems of saturation and pollution in urban transport services run on the surface, it was planned in the 1970s to lay underground metro railway lines starting with Calcutta (Kolkata). Trial runs took place from 1984 onwards and the first metro line from Tollygunge to Dum Dum (16.45 kms) was inaugurated on Sept. 27 1995. The project took two decades to complete due to heavy tunnelling involved and as a first experience for Indian engineers. The project cost financed by the Indian Railways alone escalated to Rs.16 billion. The trains run on 750V D.C. empowered through third rail. Further extensions are in progress and metro rail projects with both overground and underground tracks have since been partly completed or in progress in the cities of Delhi, Mumbai, Chennai, Bangalore, Hyderabad, Kochi and Jaipur.

Table 1: Indian Railways Zones & their Divisions with Headquarters

S. No.	Name of the Zone	Divisions
1.	Central Railway (Mumbai)	Bhusawal, Nagpur, Mumbai (CST)* Solapur* Pune^
2.	Eastern Railway (Kolkata)	Malda, Howrah, Sealdah, Asansol
3.	Northern Railway (New Delhi)	Ambala, Ferozpur, Lucknow, Moradabad, Delhi
4.	North Eastern Railway (Gorakhpur)	Lucknow, Varanasi, Izatnagar*
5.	Northeast Frontier Railway (Guwahati)	Katihar, Lumding, Tinsukhia, Alipurduar*, Rangiya^
6.	Southern Railway (Chennai)	Chennai, Madurai, Palghat, Trichy, Trivandrum
7.	South Central Railway (Secunderabad)	Secunderabad*, Hyderabad*, Guntakal*, Vijayawada*, Guntur^, Nanded^
8.	South Eastern Railway (Kolkata)	Kharagpur, Chakradharpur*, Adra*, Ranchi^
9.	Western Railway (Mumbai)	Bhavnagar, Mumbai Central, Ratlam*, Rajkot*, Vadodara*, Ahmedabad^
10.	East Central Railway (Hajipur)**	Danapur, Dhanbad, Sonepur, Mughalsarai, Samastipur
11.	East Coast Railway (Bhubaneswar)^	Khurda Road, Waltair, Sambalpur
12.	North Central Railway (Allahabad)^	Allahabad*, Jhansi*, Agra^
13.	North Western Railway (Jaipur)**	Bikaner*, Jodhpur, Jaipur*, Ajmer*
14.	South East Central Railway (Bilaspur)^	Nagpur, Bilaspur*, Raipur^
15.	South Western Railway (Hubli)^	Bangalore, Mysore, Hubli*
16.	West Central Railway (Jabalpur)^	Jabalpur, Bhopal, Kota*

*   –   Reorganised Divisions w.e.f. 1.4.2003
** –   New Zones operationalised on 1.10.2002
^   –   New Zones/Divisions operationalised on 1.4.2003

The fastest progress has been achieved in Delhi where 188 km of metro rail was put into operation in 2010 and another 130 km is in progress to be completed shortly. The track is mostly elevated, the underground portion being less, thus reducing the cost of installation. Now, the funding of such projects is shared by the Railways and State Governments and a good percentage of revenue is obtained by property development such as construction of commercial spaces over the stations, advertising etc. The gauge of the metro lines is chosen as 4 feet 8½ inches or standard gauge on account of economic considerations and the easy availability of rolling stock. The cost of elevated structures and underground tunnelling is much cheaper and the rolling stock can be picked up from the readily available inventories. However, interconnection between the metro rail and the suburban system is not possible but can be obviated by providing easy transit points.

1.7 Training of staff

Utmost importance has been given to training of railway officers and other staff and post 1947, seven centralised training institutes have been established to train officers in the four engineering disciplines, transport management, railway protection force and overall management and interdisciplinary coordination. Of interest to Signal & Telecommunication engineers is the Indian Railways Institute of Signal Engineering & Telecommunications (IRISET) established at Secunderabad in 1957. It trains both probationary officer engineers and trainee junior engineers and conducts refresher courses for serving engineers. It has trained also staff from some African and Asian railways. Working models of the electronic interlocking systems in addition to others are available for training.

1.8 Corporations etc under Indian Railway management

About thirteen corporations and limited companies have been established under Indian Railway management for some specialised activities such as managing container traffic, managing IT (Information Technology) activities of passenger reservation and e-ticketing, offering technical and economic consultancy for new lines in India and abroad, constructing railway lines in the country and abroad and many others. Of interest to signal and telecommunication engineers is the Railtel Corporation of India Ltd which was established in the year 2000. This organisation has built an Optic Fibre Communication(OFC) network along the railway track covering the whole of India, reaching about 70% of the population. Till now, 46888 route kms of OF cable has been laid and further 7201 kms will be laid in future. A robust data network on carrier class MPLS routers with multiple protocols has been established. Data centres have been established at two cities namely Secunderabad and Gurugram near Delhi. The network can cater for cloud computing requirements and retail broadband service for the public in addition to its function of providing voice, data and video services to the railway personnel.

1.8.1 Konkan Railway Corporation

The Konkan Railway Corporation was formed in 1989 with funding from four state governments in addition to railways. It was to construct a difficult railway line on the picturesque Western coast over rocky mountains and myriad rivers for a length of 760 km. This is a non-electrified single line connecting the missing link between Mangalore and Mumbai cutting short the travel distance – a person from Mangalore wanting to travel to Mumbai had to pass through Madras on the BG line as the other links were on MG till the conversions following 'unigauge policy' took shape. The first train on this railway was flagged off on Jan. 26 1998.

1.9 Present Scenario

As of now, Indian Railways are the fourth largest in the world, after the rail networks in U.S.A. Russia and China. This system runs 12000 passenger trains per day carrying over 23 million passengers connecting 8000 stations. More than 7000 freight trains are run per day carrying about

3 million tonnes of freight. The network with route kms of 65, 806 (as on 31.3.2014) and running track kms of 89, 957 is highly saturated and running additional trains to meet the needs of the public and business customers has become impossible without affecting quality of service. The passenger fares being low due to social reasons, the passenger traffic is subsidised by freight traffic. The freight rates are high, but there is a limit to the rise in freight rates, as that will divert more freight to the roads. Augmenting the line capacity by doubling/tripling/quadrupling, provision of automatic signalling, building of new lines etc. is urgently required but funding is a problem as a huge amount of Rs. 2000 billion may be required. The Ministry of Railways is exploring various methods of funding, such as funding by state governments, funding by industries such as mineral, coal and cement companies, funding by port development firms, BOT (Build, Operate & Transfer) method, also domestic and Foreign Direct Investment (FDI) in certain listed projects etc. It is expected that some of these methods will help in bolstering the capacity.

References

[1] B.V.L. Narayana – "Indian Railways – an introduction" – a power point presentation (2010) – Railway Staff College Baroda.

[2] Govind Ballabh – "A strategy for urban transport development in Delhi"–Rail Transport Journal – April – June 2010, pp. 30–40.

[3] G. Ballabh – "India's need to plan for massive urbanization" – Rail Transport Journal – Jan – March 2011, pp. 25–34.

[4] http://www.dna.com

[5] http://www.irfca.com

[6] R.M. Raina – "Funding of Metros through property development" – Rail Transport Journal – Jan – June 2012, pp. 29–33.

[7] Sopan Kasinath – "Konkan Railway Corporation" – Rail Transport Journal – April – June 2011, pp. 39–43

[8] White paper – "Indian Railways – Lifeline of the nation" – issued by the Ministry of Railways – Feb. 2015.

## CHAPTER 2

EVOLUTION OF SIGNALLING ON INDIAN RAILWAYS

2.1 Introduction

Signalling has evolved gradually on the railways with the needs of traffic, also for ensuring safety and a desirable speed. The rudimentary signalling required initially, was looked after by the permanent way engineers who laid the tracks. Signal engineering has developed now into a specialized subject, more so, with the introduction of electronics. Signalling, rather, mechanical signalling is still taught as a part of railway engineering in the civil engineering curriculum of Indian universities. There is no university yet in India which offers a railway signal engineering course whereas a few universities in Europe are now offering. It is therefore imperative that the railways themselves train the staff in the theoretical and practical aspects of signal engineering which the Indian Railways have largely fulfilled through their institutes.

2.2 Early signalling

In the early years, the signals were either revolving discs or arms with separate spectacles for light. There was no interlocking between points and signals. The signalling at way stations comprised provision of a 'main signal' in front of Station Master's (SM's) office and an outer signal in each direction without any interlocking. This method of signalling was introduced on the first railway opened on April 16 1853 from Bombay to Thane and further extensions. Later, semaphore signals with combined spectacles were installed, interlocking through keys was introduced, with the keys carried by porters between SM and the point location. Block working was through written instructions and crude communication in the beginning.

2.2.1 In 1878–79, a system of indirect interlocking was introduced at six stations between Lonavla and Poona on the GIPR. It was designed by Sir George Berkley and Mr. Wilson Bell in collaboration with M/s Saxby and Farmer of London. The installation was done with the help of a construction foreman sent from England. In 1894 Mr. G.H. List and Mr. A. Morse, engineers of North Western Railway installed a relatively advanced system of interlocking at 23 single line crossing stations between Ghaziabad and Peshawar (a part of it is now in Pakistan). Signal engineers were appointed only in the last decade of the 19ᵗʰ century, examples being Mr. J.W. Pointer of M.&S.M. Railway (Madras as headquarters) who took charge in Oct. 1893 and Mr. Dutton of E.I.R. (Calcutta as headquarters) in close proximity. The development of signalling was still not fast as seen from the position in January 1899 when there were only seven stations on E.I.R. which had some form of interlocking and overall only 13 stations out of 31 stations had cabin interlocking. The cabin interlocking comprised open elevated cabins at either end of a station to operate points and indirect interlocking through keys. This form of signaling was designed and installed by M/s. Saxby & Farmer of England. In 1904, the interlocking through keys was standardized in all railways through the efforts of Mr. Hodson, Director of Railway Construction. In 1905, a refinement of transmitting keys to SM's control electrically instead of manually was introduced. This electrical transmission was designed by Sir Lawrence Hepper, Deputy General Manager of GIPR. His name is still used to indicate this method of transmission. Interlocking the advanced starter with the block instrument and Switch and Lock Movement(SLM) at the points were gradually introduced.

2.2.2 Two Cabin Interlocking

From 1905 onwards, efforts were made to improve the standard of interlocking for higher speeds and the two cabin scheme was adopted for direct control of points and signals at either end of the yard. Operation of points and locking of facing points by rodding, mechanical detection of facing points, single wire operation of lower quadrant signals, SM's control through slide control devices etc. were introduced gradually on the mainline BG railways and by 1912, large scale cabin interlocking was completed between Bombay and Delhi on the GIPR. Similar schemes were completed on BBCI, EIR and M&SMR with local variations.

2.2.3 Double Wire Signalling

In 1920, Mr. Baker, signal engineer of Assam-Bengal Railway (present North Frontier Railway) introduced double wire signalling for operation of signals and points from a central place at single line crossing stations on that railway. As the range of operation of double wire system was higher, this proved economical and was particularly suitable for MAUQ (Multiple Aspect Upper Quadrant) signalling. This technology was originally developed in Germany and from 1950 onwards this method of signalling was adopted in many other railways, while installing MAUQ in replacement of wornout signalling.

2.2.4 Block working

In the 1940s, ball token instruments were installed on GIPR single lines due to the efforts of Mr. Neale, an engineer on that railway who developed the prototype. About the same time on double lines on EIR 'Carsen' instruments developed by Cargil and Sen Gupta were installed. Some other block instruments imported such as Bald's, Tyer's, Sykes' and SGE etc. were installed in many other railways. The SGE (Siemens and General Electric) instrument for double line is now commonly used as it is of simple design and is manufactured in railway workshops.

2.2.5 Automatic Signalling

Automatic signalling with multiple aspect colour light signals was introduced for the first time in 1925 on the harbour branch of GIPR. Later in 1928, this signalling was extended from Bombay to Byculla. Similar signalling was introduced by BBCI in its suburban sections extensively in the 1930s when it was found that the visibility of semaphore signals was obstructed due to the installation of catenary for electrification. To distinguish automatic signals and other signals and to make use of the 'stop and proceed' rule "A" marker sign was adopted in Jan. 1936. This idea was attributed to Mr.C.E. Davies, then Deputy Chief Engineer of BBCI.

2.3 Advancements in Signalling post-1947

2.3.1 Tokenless Block Working

To enhance track capacity to the fullest on single line in the steel and coal belt of India namely the eastern and mideastern part of India, tokenless block working was introduced on the Nargundy – Khurda Road section of South Eastern Railway in 1957. Japanese Kyosan and Daido instruments were used initially, later pushbutton tokenless instruments manufactured in signal and telecom workshops of Southern Railway at Podanur have been adopted for use on all the railways.

2.3.2 Route Relay Interlocking

To improve on the working of mechanical lever frames which were prone to failures, miscreant activity and operation of which was time consuming in setting points and signals, Route Relay Interlocking (RRI) system was introduced for the first time at Churchgate on Western Railway in 1958, followed by Kurla on Central Railway in 1959. This system involves just pressing two buttons at the entrance and exit, enabling setting of the route, moving the points and switching the signals automatically. The Madras Basin Bridge Junction on Southern Railway was provided with RRI in Oct. 1961 along with remote control of RRIs at three adjacent stations on three converging lines namely Vyasarpady, Korukkupet and Washermanpet from Basin Bridge.

2.3.3 Centralised Traffic Control

Centralised Traffic Control (CTC) with remote control of points and signals from a central place usually the Headquarters of the Railway or Division, full track circuiting of the yards etc. was first installed in 1966 on the North-Eastern Railway between Gorakhpur and Chapra and on North-East Frontier Railway in 1968. This was mostly with imported equipment. An indigenously developed CTC was introduced on Madras-Tambaram section of Southern Railway in 1970.

These were abandoned after a few years due to non-technical factors such as theft of cable and problems in meeting the Indian operating conditions.

2.3.4 Axle Counters

Axle counters were introduced progressively from 1972 onwards on Indian Railways, in place of track circuits, eliminating the requirement of insulated joints and wooden or concrete sleepers. The technology was originally imported from Germany but later adopted for local conditions, including digitalization of the circuitry. Axle counters have also been used for block clearance and last vehicle proving as there is no limit to the distance between the axle counter reading points.

2.3.5 Automatic Warning System

Automatic Warning System (AWS) was first installed in 1972 on Eastern Railway (ER) between Gaya and Mughalsarai and on Howrah-Burdwan section, later on Western Railway in 1987 between Churchgate and Virar, work on Central Railway (CR) suburban section soon followed. The installations on ER were experiencing frequent thefts of track magnets, hence were made inoperative. The systems on suburban sections of WR and CR are in continuous operation. The AWS controls speed of train ahead of 'yellow' and 'red' aspects and applies brakes automatically in case driver disregards a 'red' signal. Train protection and warning system (TPWS), a variant of AWS having functionality of European Train Control System (level 1) has now been installed recently on 250 routekms. This is Euro balise based and has features of automatic speed control and prevention of signal passing at danger. More lines are likely to be equipped with this system.

2.4 New age signalling with computerization

2.4.1 Electronic Interlocking

An experimental Electronic Interlocking (EI) system was installed at Srirangam of Southern Railway (SR) in 1987. This was supplied by the Union Switch and Signal (U.S&S) company of USA and of Microlok I version. The first installation for regular operation was at Kavali of South Central Railway (SCR) on July 10 1994. This was supplied by U.S & S (Microlok I) and was replaced by Microlok II, an improved version, on May 16 2008. From then onwards till now, more than 800 EI systems of various makes have been installed on Indian Railways, due to the inherent advantages of EI in remodelling, operation and maintenance. Till the year 2015, about 15% of the interlocked yards were equipped with EI, a majority of them still relay based. Overaged relay systems are likely to be replaced by EI in future.

2.4.2 Anti-Collision Device

An Anti-Collision Device (ACD) indigenously developed by Konkan Railway (KR) has been installed on 1800 kms of BG section of North Fronter Railway (NFR) in June 2007. This device is microprocessor based, is designed to prevent head-on, side and rear-end collisions at high speed. It gets location information from GPS satellites and communicates with other ACDs fitted in locos, stations and level crossing (LC) gates etc. in the radio band. Automatic braking is activated when an obstruction is found by the intelligence gathered by the microprocessor system. The device also provides audio visual warning to LC gates both manned and unmanned while approaching. More such devices are planned for installation in other sections.

2.4.3 Train Management System

Train Management System (TMS), a modern version of CTC has been installed on the suburban sections of Western Railway (WR) – Churchgate to Virar and Central Railway (CR) – Mumbai to Kalyan in the years 2003 to 2008. The movements of all trains on these sections are displayed on a wall opposite to the positions of the traffic controllers on a real time basis. The status of points, signals, track circuits, route settings etc. are fed regularly to the TMS computer system, as well as the train identification information, from the train originating stations. There is provision to remotely activate the points, signals and routes by the traffic controller in addition to the station Master. Real time updates on train movements are fed to Public Address (PA) systems at all the stations for the information of passengers. As a part of the controller's duties, the train graphs, punctuality reports, rake and crew links etc. are generated more easily with the aid of the TMS computer system.

2.4.4 Data Loggers

Microprocessor based data loggers are being installed at each interlocked yard to aid in maintenance, are also networked to feed information to a central monitoring system. These help to detect unusual happenings such as drivers passing signal at danger, driver passing a turnout at excessive speed, equipment failing in unsafe condition etc.

References

[1] A. Vijayvargiya – "Role of signalling for safety…" – a power point presentation (2008) – Railway Staff College Baroda (Vadodara).

[2] B. Tayal and N.S. Deo – "A cost effective system for railway level crossing protection" – a power point presentation (2008) – Konkan Railway India.

[3] Chandrika Prasad – "End of the search for a suitable automatic warning system for Indian Railways" – Journal of the Institution of Railway Signal & Telecom. Engineers (IRSTE) – India – Dec. 1997, pp. 5–12.

[4] http://www.irfca.com/tms

[5] Indian Railways Knowledge Portal – "Development of Signalling on Indian Railways" – a web document (2016).

[6] Konkan Railway – "ACD network – a train collision prevention system…" a power point presentation (2007).

[7] Silver Jubilee Souvenir – "Through the mist of antiquity" – Indian Railways Institute of Signal Engineering and Telecommunications – Nov. 1982, pp. 36–39.

[8] White Paper – "Indian Railways – Lifeline of the nation" – issued by the Ministry of Railways – Feb. 2015.

## CHAPTER 3

INTRODUCTION OF ELECTRONIC INTERLOCKING

3.1 World Railways

After the advent of the microprocessor in the 1970s, there was tremendous interest in introducing microprocessors in the control of industrial processes both in the academia and the industry. Short-term and part-time courses were introduced in the universities on the subject of microprocessors and applications in the late 1970s. The World Railways also turned their attention to this new field from 1976 onwards due to the inherent advantages in the introduction of microcomputers. First installation of Solid State Interlocking (SSI) at Lemington Spa on British Rail was inaugurated on Sept. 8 1985. Similar installations were being commissioned by other European Railways. The American Railways had also experimented with Vital Processor Interlocking (VPI) systems with single processor from 1975 onwards with the aid of signalling firms, namely Union Switch & Signal and General Railway Signal Co. The first installation of Microlok (US&S) was at Esplen of Conrail (Aug 15 1985).

3.1.1 Developments on Indian Railways

In India, the Research Design & Standards Organization (RDSO) of Indian Railways (IR) took up the development of SSI in collaboration with the Indian Institute of Technology (IIT) Delhi in July 1983. A prototype was installed at Brar Square station of Northern Railway (NR) in parallel with the relay interlocking system in 1987. Electronic Interlocking (EI) systems based on this design were installed for regular operation at three wayside stations progressively between 1995 and 1998, through a contract with three Indian firms.

3.1.2 Some experiments were also conducted at the Indian Railways Institute of Signal Engineering of Telecommunications (IRISET) at Secunderabad in 1982–83 at the instance of the author. A model of a circular railway was chosen and one of the stations was controlled by a microprocessor. The interlocking software was based on an assembly language program converted to machine code and loaded manually. This program was stored in an EPROM after complete check of the functions. The details of the experiments were published in [4] and [5], more details are given in [6]. The concept of safety by system error detection and achieving lower accident probability were illustrated. Some of the relevant details are included in Appendix A for ready reference while discussing safety and reliability in the subsequent chapters.

3.1.3 An EI system of the U.S &S make (Microlok I) was installed in parallel with the relay interlocking system as an experiment at Srirangam of Southern Railway (SR) in 1987, later it was put in series and in 1989 it was tried as a stand alone system which was sucessful. It was later taken out and the relay system retained. The EI system was approved by the Indian Railway Board for installation at wayside stations from then onwards taking into consideration the advantages of introducing electronics, which are enumerated below.

3.2 Advantages of Electronic Interlocking

(i) The main advantage of electronic interlocking is the ease in doing alterations while remodelling a yard where only the application software needs to be altered. This reduces the non-interlocking period to a few hours.

(ii) Building costs will be reduced as the space for electromechanical relays required for interlocking is saved by the introduction of less bulky PCB (Printed Circuit Board) racks. It is estimated there will be 25% reduction in space.

(iii) With serial data transmission and use of OFC (Optic Fibre Cable), bulky copper cables are eliminated, resulting in savings.

(iv) Power consumption is reduced due to the preponderance of electronics

(v) With the provision of self-diagnostic features and data loggers, duration of failures can be reduced, resulting in less maintenance staff.

(vi) The cost of EI system in India, though high in the initial stages due to the import content, is considerably reduced due to progressive indigenisation and further reduction is likely.

3.3 Installation of Microlok I of U.S.&S.

Based on the approval of the Railway Board, South Central Railway (SCR) entered into a contract with Railway Products (India) Ltd (RPIL) for supply and installation of Microlok I (US&S) at three stations on Vijayawada – Gudur section in 1991. The first installation of the system at Kavali was inaugurated on July 7 1994. This is reckoned as the first installation on IR for regular operation as the one at Srirangam was removed after the experimentation. The systems at the other two stations Srivenkateswarapalem and Manubolu were inaugurated on Oct. 15 1994 and Nov. 15 1994 respectively. These were however replaced by the Microlok II systems in the year 2008. Subsequent to this, more than 800 EI systems of various makes have been installed on IR. The predominant system is still the Microlok II, now supplied by M/s Ansaldo of the Hitachi group who have acquired the American firm U.S. & S. The Microlok I was the older version in which the non-vital controller or genisys was separated and the I/O capacity was limited. The later version of this, called Microlok Plus in U.S.A, termed Microlok II with some more modifications in India is based on Motorola 68332 microcontroller and has a higher speed of 21MHZ. It has expanded diagnostics and an integral event recorder. The non-vital processing with inputs from control panel etc is carried out by the single microcontroller instead of housing a separate controller as in genisys. 578 such systems have been installed till now.

3.3.1 Microlok II of M/s Ansaldo STS (India) Ltd

The basic system and the processing of vital logic in Microlok II are illustrated in figures 3.1 and 3.2. For safety assurance, a number of self testing and diagnostic programs are included in the single microcontroller. Programs to check CPU, registers, memory etc are present along with hierarchical routines i.e. one software program leads to another software program and so on. The two important routines are (i) dual path processing wherein the safety logic is processed twice and processed further only if there is correspondence (ii) double store, wherein all data is stored in two different RAM chips and the data accessed from each clip is processed by two different softwares, one normal and the second with D'Morgan's equivalent and the results should correspond. If these routines correspond, vital clock signal is sent to a power circuit which keeps the power in position. If not, the power to output circuits is switched off. Also vital kill circuitry goes into operation i.e. the power to the microcontroller itself is cut off.

3.3.2 In addition, a closed loop principle is used to monitor inputs and outputs. The inputs are momentarily de-energised for a millisecond and monitored by the microprocessor, confirming the positions. Similarly the outputs are monitored to see that they correspond to the output drive signals.

3.3.3 Application Software

The software embedded is divided into two parts (1) Executive Software and (2) Application Software. The executive software pertains to the normal operation of the processor including the operating system and is embedded in an EPROM supplied by the manufacturer. The application software is connected to the vital logic dependent on the interlocking requirements of the station yard. The conventional relay wiring diagram is converted to a boolean expression and loaded into the compiler supplied by the firm. The output which is equivalent to the '.exe' code is stored in an EEPROM. This software only needs alteration if the yard is to be remodelled and can be done by the user without any external assistance.

3.3.4 Standby modes

Microlok II being a single processor system, there is provision for either a warm or hot standby. The requirement now is a hot standby, to avoid the time gap for switching on the standby and warming up which takes 3 to 5 minutes. The common mode failure in a hot standby system cannot be avoided but the probability of such failure is quite low. In coastal areas, which are prone to heavy lightning, warm standby is preferred as there is a chance of the hot standby failing simultaneously. Additional circuits are required to keep the system fault free in both warm and hot standby operating modes.

3.3.5 Communication based remote operation

Master-slave system can be adopted for saving on multicore copper cables for a station yard wherein slave microlok units or object controllers can be installed at each end of the yard nearer the points.

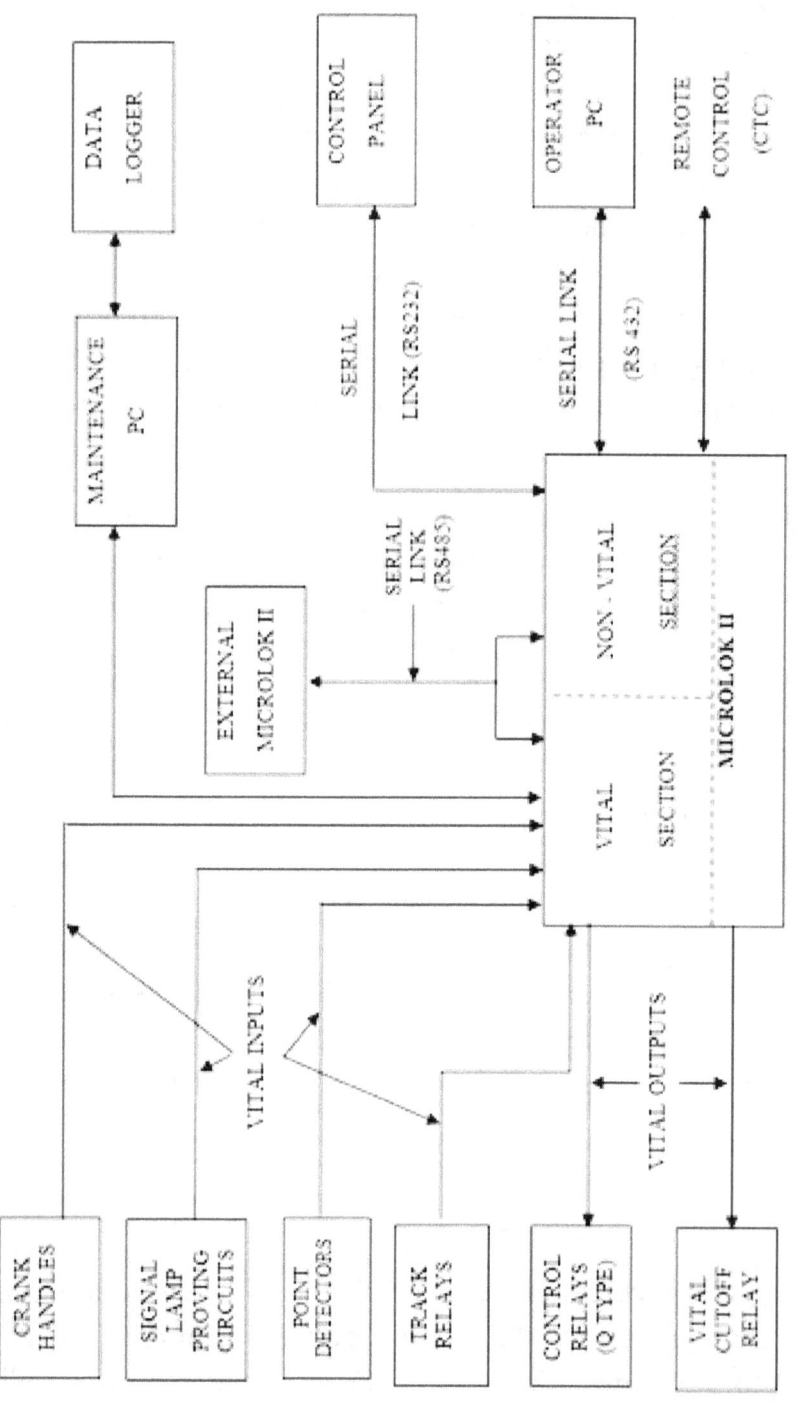

Fig.3.1 Basic Microlok II System (*simplified*)

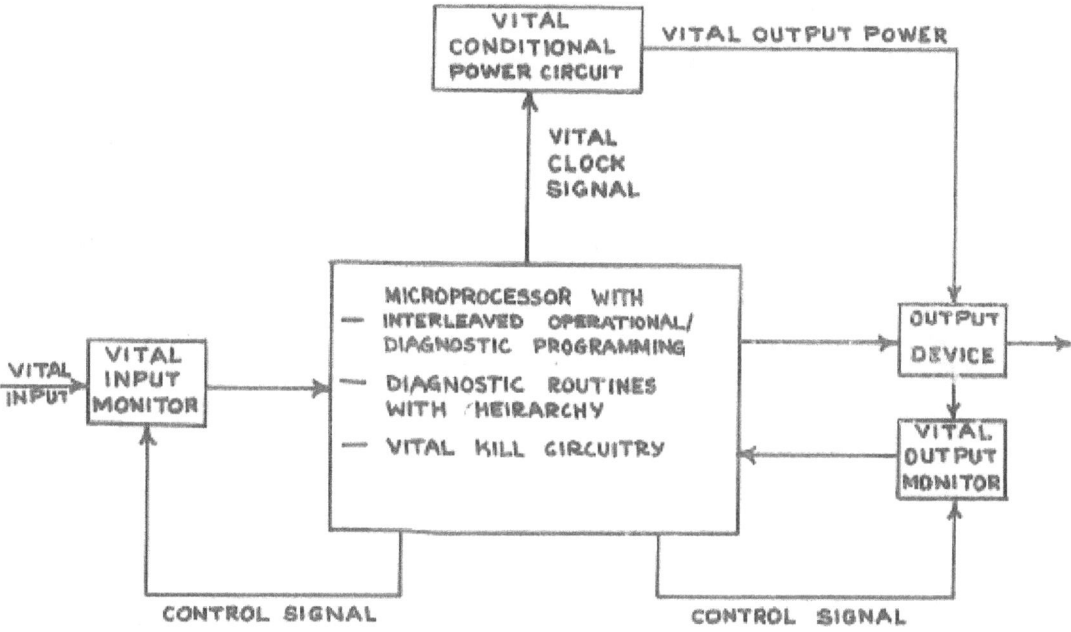

Figure 3.2 Processing of Vital Logic (Microlok II)

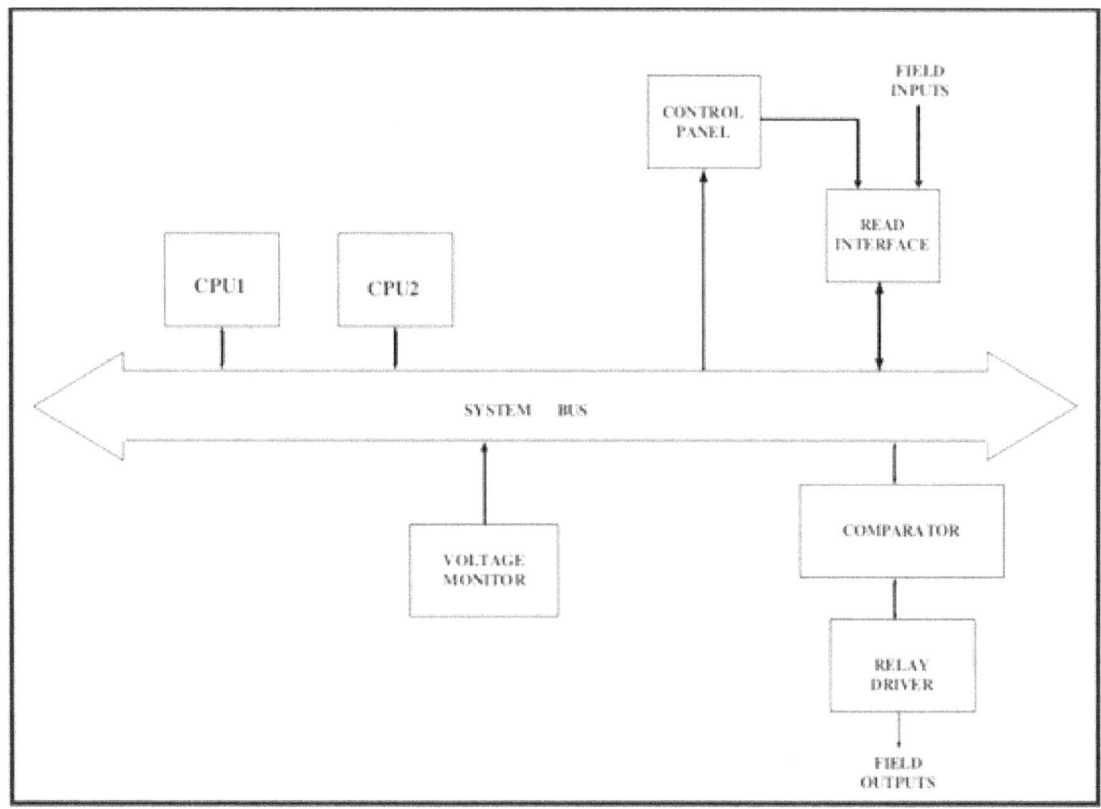

Figure 3.3 Block Diagram of E.I. (RDSO) Mark II

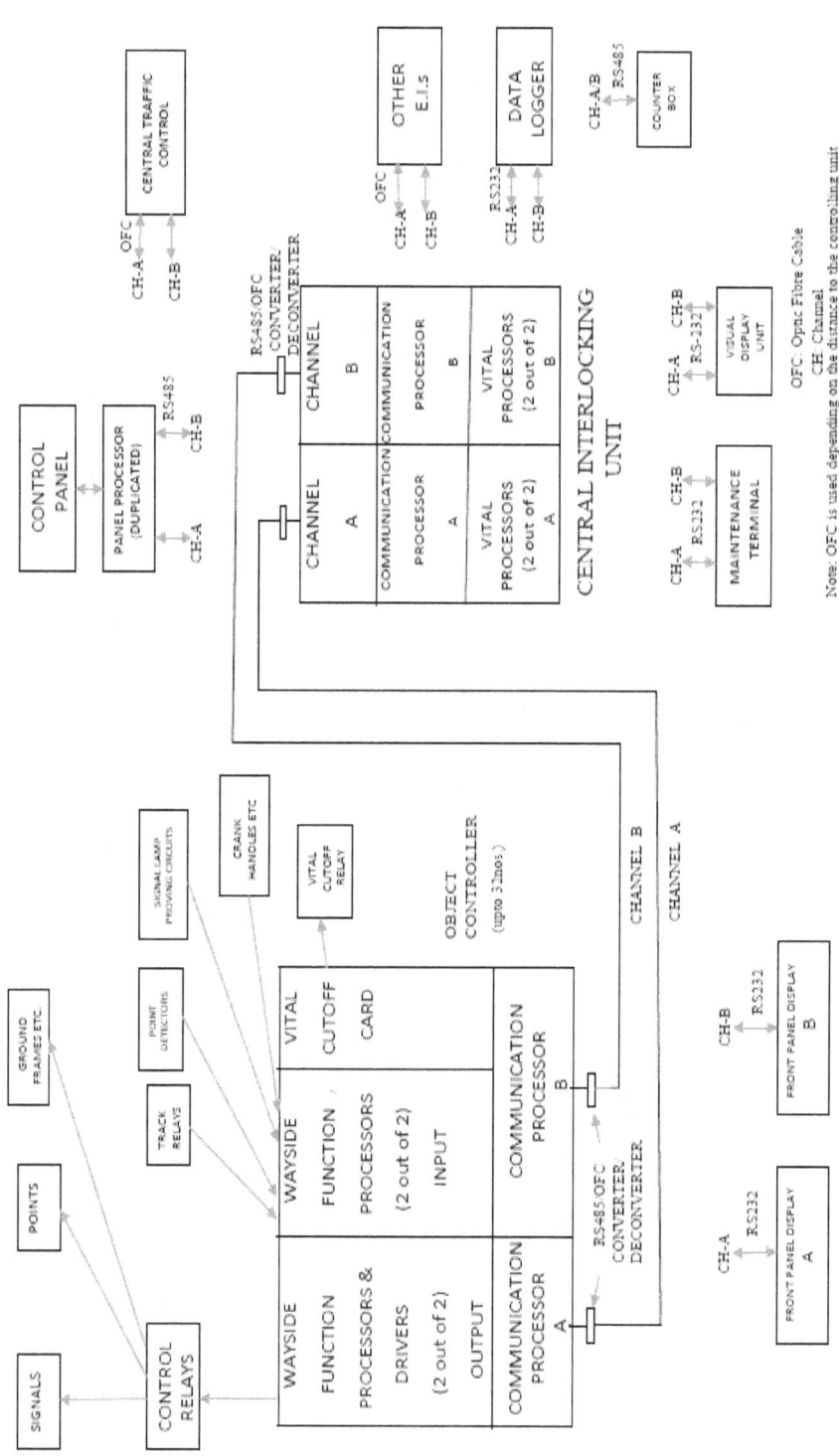

Figure 3.4 Block diagram of Medha Electronic Interlocking (MEI633) (simplified)

The communication between the distant microloks can be through traditional serial links or through optic fibre cable, with suitable interfacing modules and communication boards at the card level. Ethernet based networks are adopted for speedy communication with TCP/IP (Transmission Control Protocol/Internet Protocol). Some stations on the East Coast Railway are equipped with such systems.

3.4 R.D.S.O. type Mark II

Experimental installations fabricated according to the design of R.D.S.O (Research Design and Standards Organisation) of Indian Railways were installed

(i) At Dushkheda of Central Railway in Oct. 1995 – the manufacturer being M/s DCM Ltd.,

(ii) At Bhadli of Central Railway in March 1997 done by M/s Central Electronics Ltd

(iii) At Uppugundur of South Central Railway in Apr 1998 done by M/s Railway Products India Limited.

3.4.1 The block diagram of the system is given in Fig 3.3. Two identical CPUs (8086) are employed and two identical softwares in different time slots are embedded. A failsafe comparator compares the outputs and only if identical, the external output to relay driver is signalled. The salient features are

(i) Individual hardware modules have been assigned unique signatures which are verified by the microprocessors in every cycle to ensure that the correct module is accessed.

(ii) The address of I/O port accessed by the microprocessor is latched on the hardware module so that it can be verified at a later stage.

(iii) Each input channel is toggled to verify that the input is not stuck, to avoid 'stuck-at' faults.

(iv) Outputs generated are fed back and verified before the commands to external devices are communicated.

(v) Self-check programs to check internal hardware faults are continuously executed

(vi) Watchdog pulses are generated on completion of each self-check program and are checked.

(vii) The system is shut down in case any catastrophic failure occurs.

(viii) The voltage monitor continuously monitors the supply voltages and in case of discrepancy leads to shutdown.

3.4.2 The systems at the three stations were uninstalled after fault free service of a few years due to lack of maintenance support, also the firms could not sustain, due to lack of sufficient orders. Indigenous firms have been encouraged by the Railways to build on this expertise and one indigenous firm is able to sustain in this field. The indigenous system is described in the next paragraph.

3.5 Medha Electronic Interlocking (ME I633)

The Electronic Interlocking supplied by M/s Medha Servo Drives Pvt. Ltd of Hyderabad is designed as a multiprocessor distributed system, with the controlling processors in the Central Interlocking Unit (CIU) and slave processors in the Object Controllers (OCs) located either near or far extending over a few kilometres. A block diagram of the EI is illustrated in Fig 3.4 in a simplified form. The OC is nearer the place of operation of points and signals and there can be a saving in multi-core copper cables by using an Optic Fibre Cable (OFC) with the required interface modules.

3.5.1 Vital processing is carried out in the Central Interlocking Unit (CIU) based on the stored interlocking logic, with non-vital inputs from the control panel and/or Visual Display Unit (VDU) keyboard and vital inputs from the Object Controllers (OCs). The output data is transmitted to Control Panel/VDU, OCs, Maintenance Terminal (MT) and the Counter Box (CB). The CIU has two vital interlocking processor units connected to two communication processors which communicate through two channels (A and B-only one active) to the external devices. In case of failure of the active channel, the standby channel is activated automatically. The Object Controller has two communication processors connected to the channels A & B and are in turn connected to Wayside Function Modules (WFMs) – 8 nos at the maximum. Each WFM has two processors – master and slave – which perform the input and output functions for the field devices. The system is designed for a maximum of 32 OCs. Each WFM can have 8 functions either input or output.

3.5.2 The Panel Processors (PPs) are duplicated and connected to the two communication channels A & B. The Counter Box is used to drive counters and buzzers taking data from the active channel. It also indicates the status of vital processor units A & B and wrong side failure information. Front Panel Displays A & B display the system faults and recovery messages.

3.5.3 Processing of Vital Logic

This E.I. is basically a two out of two system with identical hardware and software and a hot standby. The vital logic processing is illustrated in Fig 3.5. The vital processor units are duplicated, also the interconnected communication processors, the communications links and interfaces and the I/O communication processors except the processors in the wayside function modules. The Control Panel processors are also duplicated. In case of any failure, the active channel is transferred to the standby channel right through the CIU, Communication links and the OC.

3.5.4 Processing of the interlocking logic is done by the dual processors in the CIU with the field inputs from OCs and operator commands from the Panel Processor (PP) and/or Visual Display Unit (VDU) keyboard. Any incongruence in the dual processors read by the supervisory processor causes interruption to the power supply to the processors. The logic is based on the application software specific to the yard stored in a Read Only Memory(ROM) or more commonly EEPROM. A subset of 'C' language as per CENELEC standard is used to develop the software. The outputs are communicated to the OC and the PP/VDU. The vital inputs are also compared by the comparator in the WFM and any dissimilarity shuts down the supply. The vital outputs are compared in the WFM and when congruent, are led to the relay driver circuits which control the signals, points etc. The power to the driver circuits is through the picked up contacts of a vital cutoff relay which drops in case of any abnormality including the incongruence in the outputs of the dual processors. The Voltage and Health Monitoring(VHM) card also aids in cutting off the supply in case of any abnormality both in CIU and OCs.

3.5.5 A normal processing cycle in this system takes 333 ms out of which 180 msec. is for interlocking logic execution. Detection of wrong side failures or inactivity and enabling a safe state is done within 2 cycles or 666 msec.

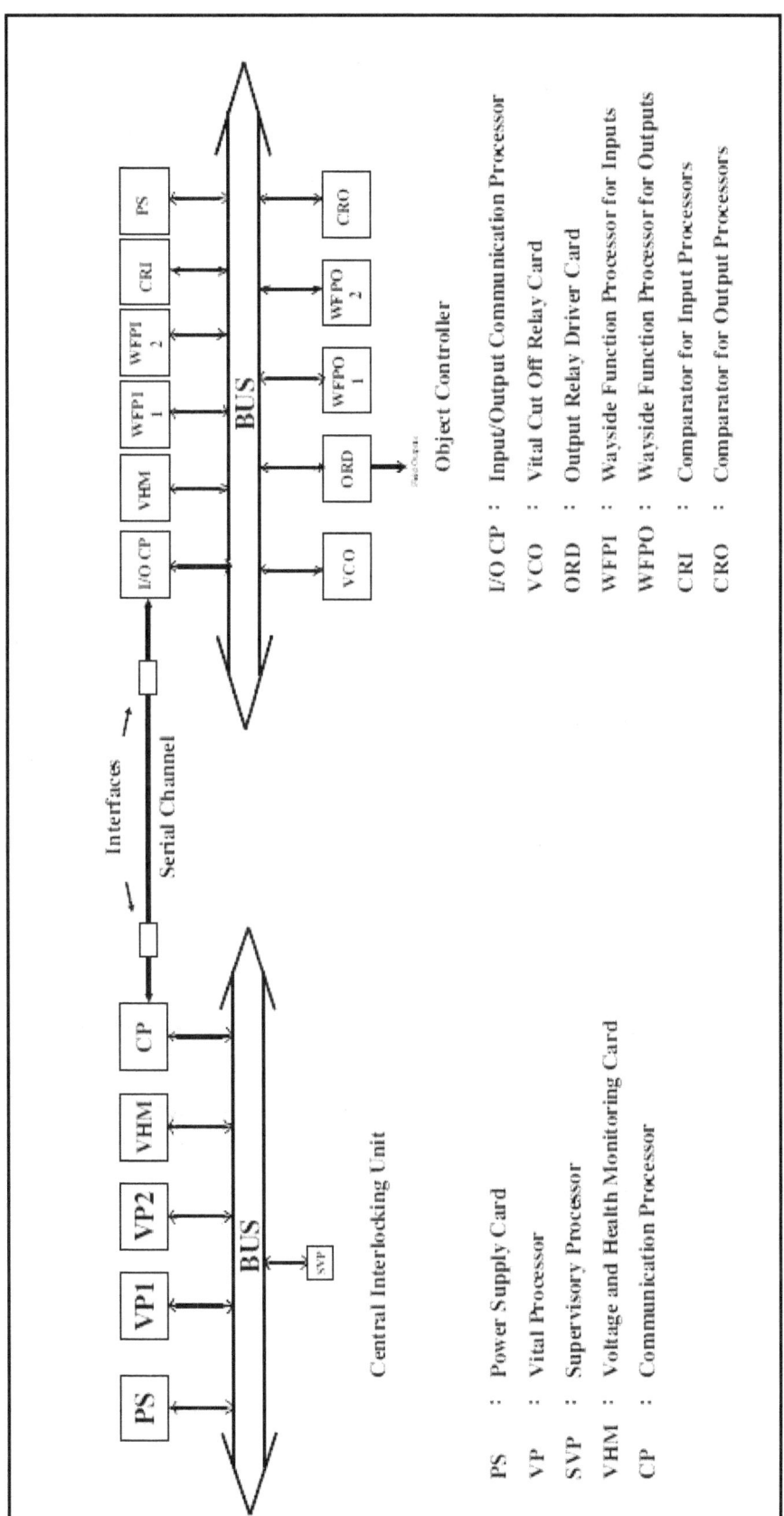

Figure 3.5 Architecture of vital logic processing in Medha E.I. (for single channel)

PS : Power Supply Card
VP : Vital Processor
SVP : Supervisory Processor
VHM : Voltage and Health Monitoring Card
CP : Communication Processor

I/O CP : Input/Output Communication Processor
VCO : Vital Cut Off Relay Card
ORD : Output Relay Driver Card
WFPI : Wayside Function Processor for Inputs
WFPO : Wayside Function Processor for Outputs
CRI : Comparator for Input Processors
CRO : Comparator for Output Processors

3.5.6 The vital variables are stored in two states, normal state and complement state and processed by reading both the states. The diagnostic and self-check programs processed by the vital processors include software timers test, program/application data memory integration tests using checksum verification, relay contacts readback and final output readback tests. Error analysis is also performed to determine the criticality of failure. A wrong side failure, namely output relay read as picked up when not required, results in cutting off supply to the cutoff relay which in turn cuts off the supply to the drive relay circuits. This is within 350 msec. If the wrong side failure persists and undesired field operation, such as, clearing a signal happens, the system is shut down after a correspondence check within 1.3 sec.

3.5.7 This EI system was approved by RDSO in 2010 and 56 installations have been commissioned so far on various zonal railways.

3.6 Westrace VLM6 – Electronic Interlocking

This system has been developed by M/s Invensys Westinghouse Australia and has now been taken over by M/s Siemens Rail Automation (P) Ltd India. WESTRACE is an acronym for WEStinghouse Train Radio Advanced Control Equipment.

3.6.1 This is a single processor system with self testing and diagnostics in the vital processor and monitoring of the health of the processor by an adjacent vital module. The block diagram of the system is given in Fig 3.6. The vital processing, based on the interlocking logic is done in the Vital Logic Module (VLM) taking vital inputs, i.e. track relays, point detection relays, signal lamp proving relays, crank handles etc. and delivering vital outputs i.e. output voltages to point control relays, signal control relays, ground frame controls etc. Similarly the non-vital inputs such as the commands from the local control panel, remote control etc are processed in the non-vital logic module and the non-vital outputs, such as, indications of signals, tracks etc are transmitted to the control panels, maintenance terminals, data loggers etc. with the required outputs from the VLM.

3.6.1.1 The Serial Communication Module (SCM) is also connected to the ethernet network for remote communication through optic fibre etc. to other Westrace systems. It is connected to the local control panel through interface modems, such as, RS-232C. Remote serial links through optic fibre cable and modems connect remote control centres, such as, Central Traffic Control (CTC). Maintenance Terminals including diagnostic fault displays etc. and data loggers are connected to the serial communication modules, also called Network Communication Diagnostic Modules (NCDM).

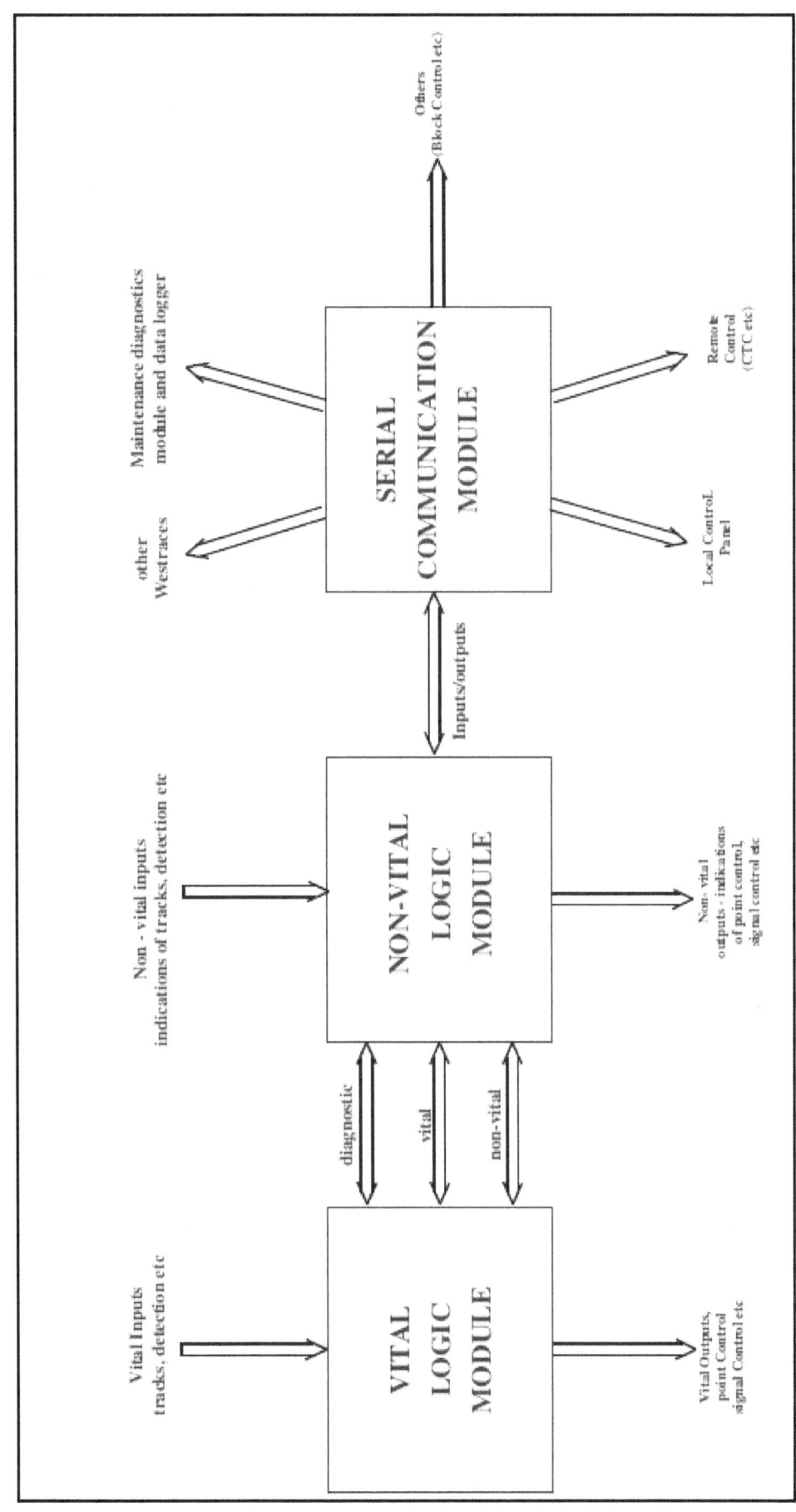

Figure 3.6 Block diagram of westrace – VLM

3.6.2 System Negation

The system reaches a safe state by means of negation of the output, in case of normal failures or abnormalities found by self testing. The Output Power Control Relay (OPCR) is energised which cuts off power to the vital function control relays, thereby shutting the system. Serial outputs are cut and serial communication modules are isolated by removing the Vital Serial Enable Voltage (VSEV). Primary negation of shutting down the Vital Logic Module is achieved by the abnormal result of self testing hardware and software within the module. Another way of negation is by monitoring of the health of a microprocessor by another microprocessor which is in the adjacent module and in case of a deficiency, the power to vital outputs is cut in the same way through OPCR and VSEV. This is called secondary negation. In practice, primary negation and secondary negation take place almost simultaneously cutting the power to vital outputs. Non-vital processing modules are not involved in the negation logic.

3.6.3 Graceful degradation

There is a provision in the system for graceful degradation, that is, if a single input or output is faulty, such links are isolated first, continuing with the rest of the system. Only if it does not work, the VLM is shut down.

3.6.4 168 systems of this type have been installed on the Indian Railways so far.

3.7 Kyosan Electronic Interlocking

The block diagram of Kyosan E.I. is given in Fig 3.9 and the architecture of vital processing is illustrated in Fig 3.8. It has a two out of two architecture with identical hardware and software. The comparator checks the outputs in every cycle and outputs positive alternations in case of congruence. These are rectified and feed a system relay. The top (or picked up) contacts of this relay are in series with the power supply input to the vital output driver circuit. In case of non-congruence, the system relay drops and cuts the power supply to the vital output drivers. The system works with a hot standby with four vital processors in all, enhancing reliability and availability. The vital outputs are transmitted to the field through optic fibre cable and appropriate serial interface and the other inputs and outputs are transmitted between the processors and the field or panel processor through the interface (IF486). Depending on the length of internal wiring, serial interface modules are employed without fibre optics, for example, wiring to the maintenance terminal, PC and panel processor etc.

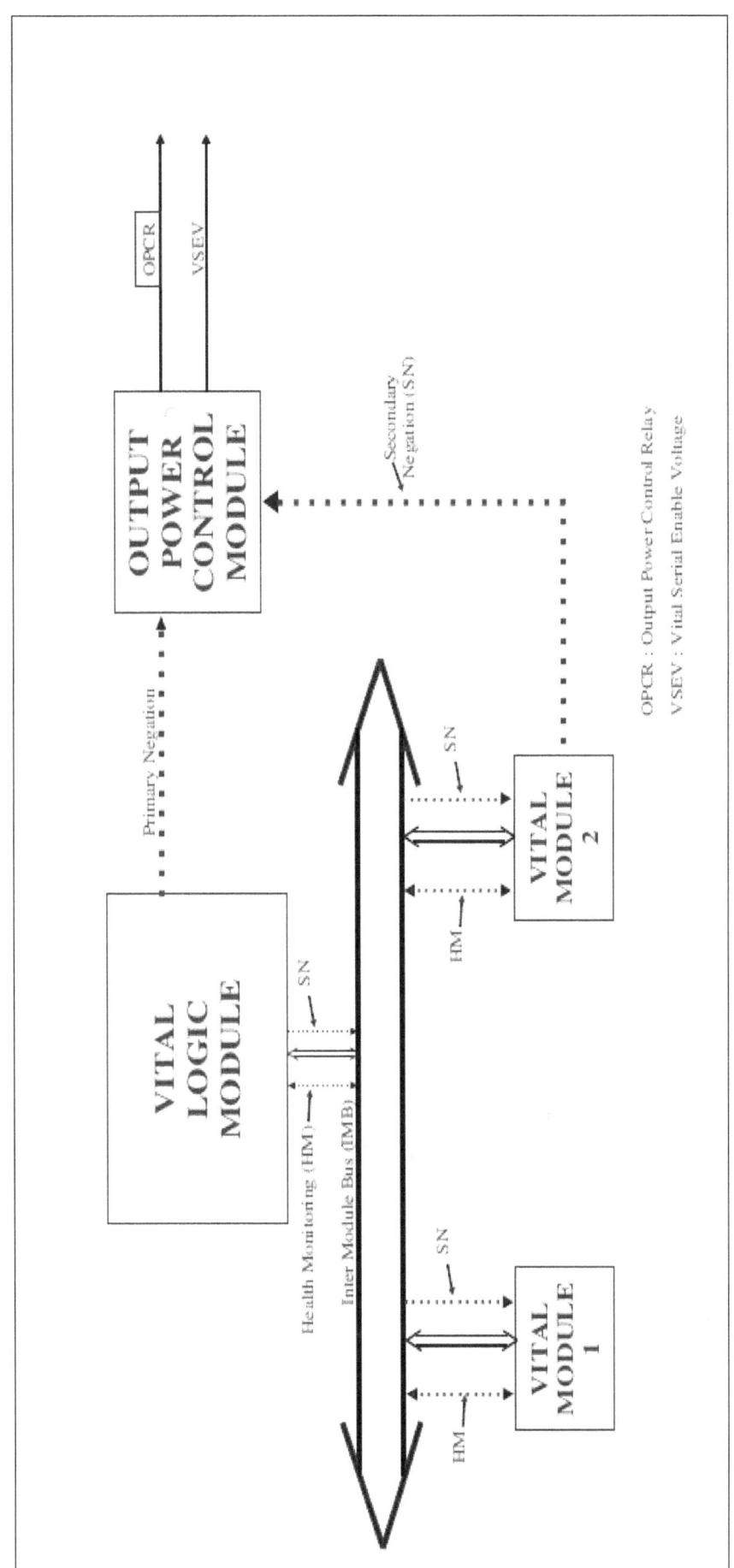

Figure 3.7 Westrace logic processing system negation

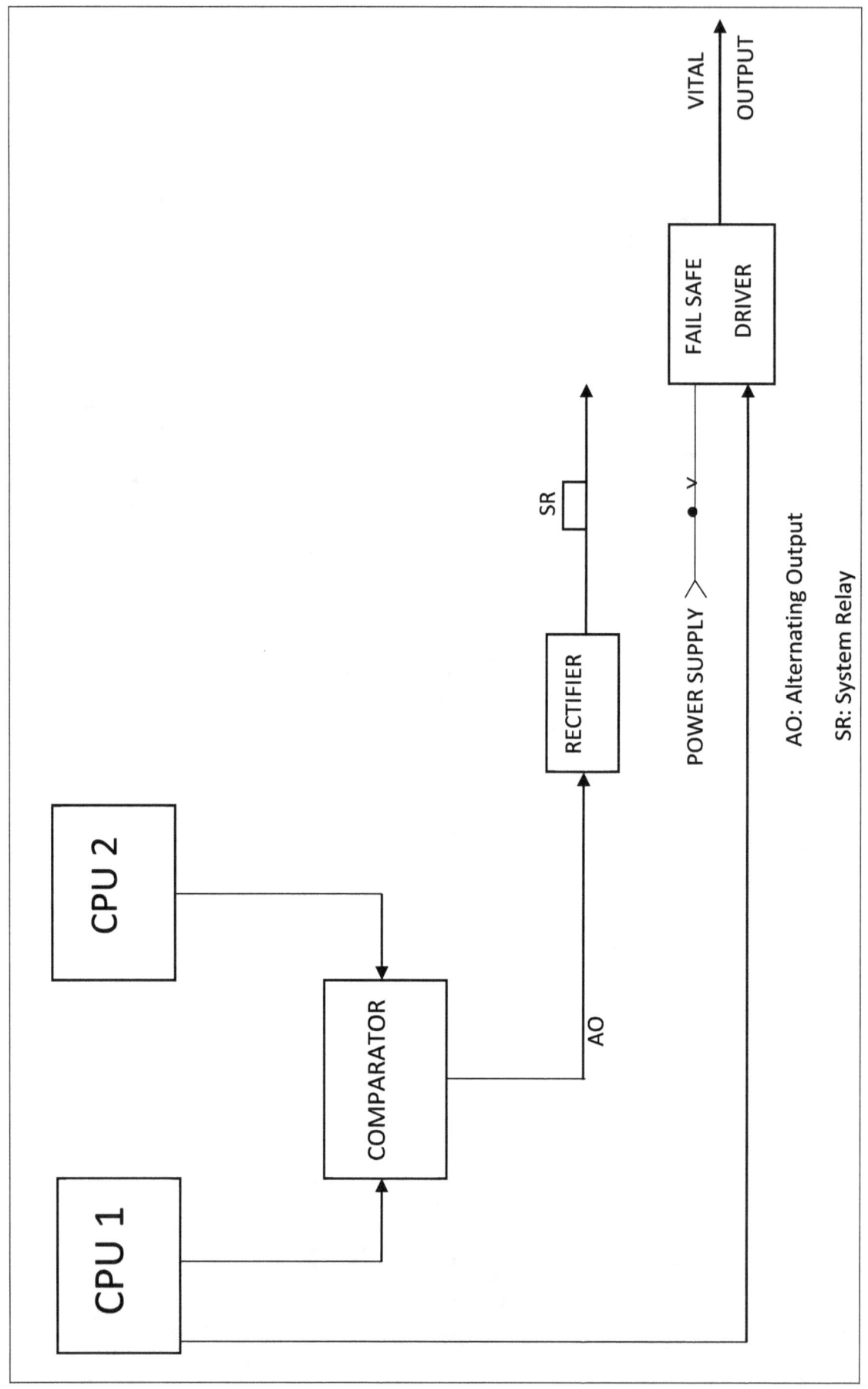

Figure 3.8 Architecture of vital processing in Kyosan E.I. (2 out of 2)

AO: Alternating Output

SR: System Relay

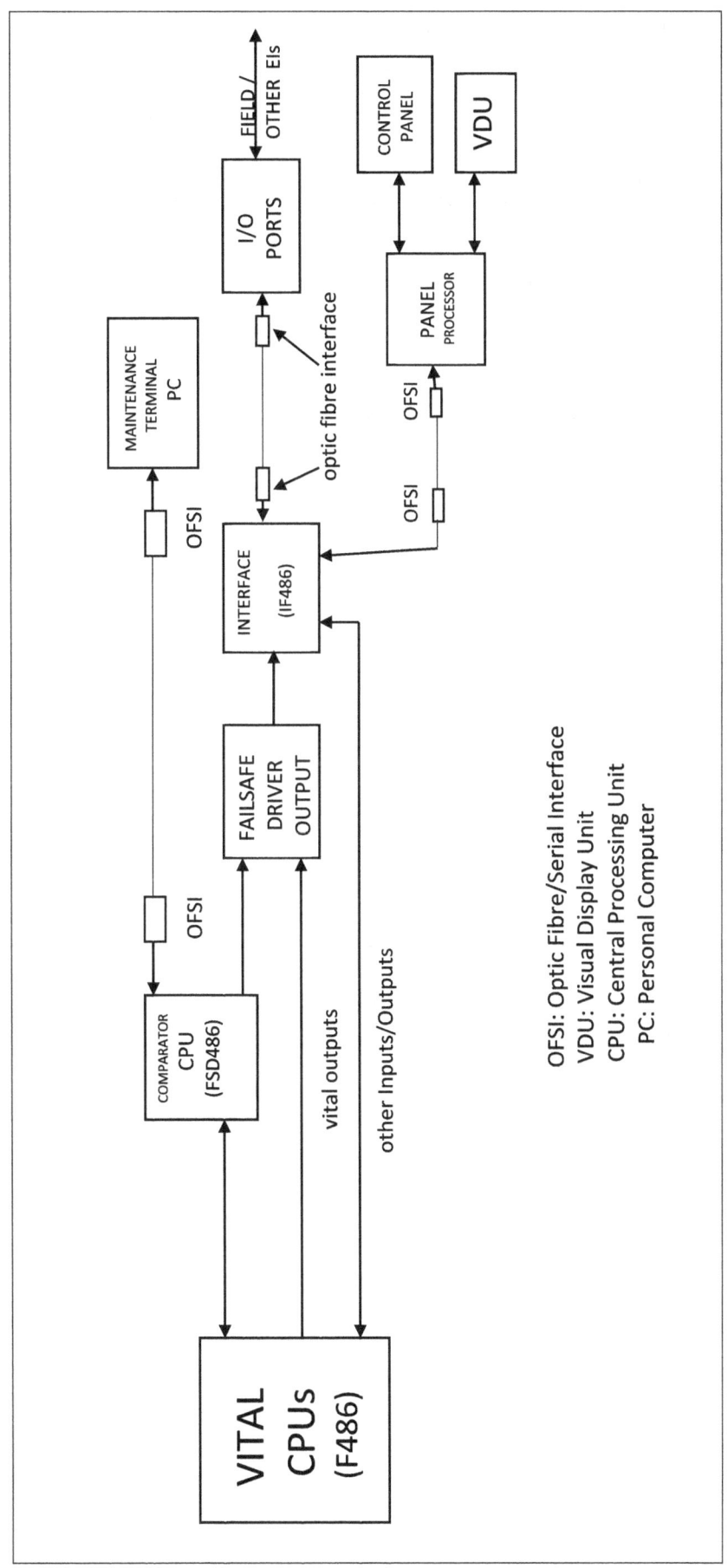

Figure 3.9 Block diagram of Kyosan E.I. (simplified)

OFSI: Optic Fibre/Serial Interface
VDU: Visual Display Unit
CPU: Central Processing Unit
PC: Personal Computer

3.7.1 The interlocking logic based on the wiring diagram is compiled in the PC through the special compiler and the output is saved in an IC/memory card manufactured by the firm. The data from the IC card is loaded into the memory of the E.I. system upon start up. Any modifications later can be easily done through the compiler and the IC card modified accordingly.

3.7.2 Only 14 systems of this type have been installed on various zonal railways in India.

3.8 Configuration of hardware and software in the EI systems on IR.

To summarise, the configuration of the microprocessor hardware and software in the EI systems on the Indian Railways is as follows:-

1	Microlok II of M/s Ansaldo	-----	single processor with diverse software
2	Medha E.I.	-----	2 out of 2 system with diverse software
3	Westrace VLM6 – E.I.	-----	single processor with diverse software
4	Kyosan E.I.	-----	2 out of 2 system with identical software
5	ESA 11 of AZD Praha*	-----	2 out of 2 system with diverse software
6	VHLC – E.I. of GE*	-----	2 out of 2 system with diverse software
7	SIMIS – S E.I.*	-----	single processor with diverse software

described in Appendix B

3.8.1 As per the regulations on IR, only hot standby is permitted to reduce the time for switch over. At present, the first four systems in the above list are approved for installation on mainline railways of IR. The other three systems listed above, have been described in Appendix B. Diverse systems have been installed on the Metro Railways or the Urban Transport System, such as Delhi Metro Railway where Alstom-ASCV-Smartlok, a single processor system with diverse software has been installed.

References

[1] Cribbens A.H. – "Solid State Interlocking (SSI): an integrated electronic signalling system for main line railways", Proc IEE Vol 134, Pt B, May 1987, pp148–158.

[2] Cribbens A.H. – "The Solid State interlocking" – Proceedings of the International Conference on "Railway Safety, Control and Automation towards the 21st century "held in London 25–27 Sept. 1984, pp24–29.

[3] D.R. Disk – "A unique application of a microprocessor to vital controls" – Proceedings of the International Conference on "Railway Safety, Control and Automation towards the 21st century" held in London, 25–27 Sept.1984, pp 97–104.

[4] V. Purnachandra Rao and P.A. Venkatachalam "Concept of safety by system error detection in solid state interlocking" – Proceedings of International Conference – "Railway Safety, Control and Automation towards 21st Century" London, 25–27 Sept. 1984, pp65–71.

[5] V. Purnachandra Rao and P.A. Venkatachalam – "Microprocessor based Railway Interlocking Control with low accident probability" – IEEE Transactions on Vehicular Technology, Aug. 1987, pp 141–147.

[6] V. Purnachandra Rao-"Studies into the application of microprocessors to control of railway signalling" Ph.D. Thesis submitted to Anna University, Madras – May 1992.

[7] V. Raghunathan – "Microlok-an introduction to SSI" – A presentation to the 5th seminar for SAG officers on "Solid State Interlocking" – 14–16 Feb.1989 IRISET Secunderabad.

[8] B.L. Robinson and C. Mokkapati-"Microlok Interlocking Controller"-Journal of the Institution of Railway Signal & Telecom. Engineers (IRSTE) India – Sept 1996 pp2–21.

[9] Signal Directorate RDSO India – "The SSI Project" issued by RDSO Lucknow, February 1995.

## CHAPTER 4

SAFETY VS. RELIABILITY

4.1 Introduction

With the extensive introduction of computer controlled systems not only on the railways but also in the fields of aerospace, defence, nuclear and chemical industries, the concept of safety through the development of highly reliable computer system (which is hardware plus software) has changed. The safety of the whole system which is not only the computer but also the personnel controlling the system, the interface to other systems and the environment, has to be considered for the prevention of hazards and accidents. This involves system safety engineering or systems engineering in general and will be considered in the subsequent chapters.

4.1.1 After the end of World War II, efforts were made to improve the reliability of electronic components and highly reliable components came out of production. A highly reliable system worked well so long it was purely analog. But with software controlled systems and digital hardware, even systems with highly reliable components started failing, resulting in hazards and accidents. To prevent frequent accidents in computer controlled systems, a lot more factors have to be taken into account other than the reliability of hardware. A lot more emphasis on the property of 'safety' is needed. When we think of safety, it should refer to system safety and not just that of the hardware. The safety of a system can be defined as "the absence of accidents, where an accident is an event involving an unplanned and unacceptable loss" as per Prof. N.G. Leveson. The system safety engineering can be defined as "a planned, disciplined and systematic approach to identifying, analysing and controlling hazards throughout the life cycle of a system in order to prevent or reduce accidents" as per Jerome Lederer. On the other hand, reliability can be defined as "the probability that an item will perform a required function without failure under stated conditions for a stated period of time" and reliability engineering is evolved "to apply engineering knowledge and specialist techniques to prevent or reduce the likelihood or frequency of failures" as per Patrick D.T.O'Connor et al.in [4]. Safety and reliability are therefore different properties to be analysed in different ways and safety is of paramount importance in dealing with the electronic safety critical systems on the railways.

4.2 Concurrent error detection system

The use of duplication, triplication of processor modules and N-modular redundancies will only increase the reliability of the system but is not likely to improve the safety unless other measures are taken. The reliability mainly refers to the hardware components and random failures and over the years highly reliable components are being produced as a result of developmental efforts by the manufacturers. While considering safety, the behaviour of the system which includes hardware and software has to be considered. To check the health of the software continuously, it has become common to insert error-checking codes or programs along with the running programs, such as parity checkers, multierror detecting codes for memories and residue codes etc for arithmetic logic units. Many more diagnostic routines are being introduced for the same purpose and the objective is to have an orderly shutdown of the system or revert to a known safe condition, upon detection of unsafe conditions. Incidentally, the reliability of the system can also be enhanced, by having a concurrent error detection system as detailed in [5]. Consider the dual redundant system with error detection as in Fig 4.1. Such systems have been termed as dynamic redundant systems by N. Storey in [3]. The processor modules M1 and M2 are connected to a validation gate through separate error detectors D_1 and D_2. The error detector hardware operates the error detection algorithm and indicates to the validation gate the presence or not of an error. The gate releases the outputs of M_1 and M_2 only if there are no errors found by D_1 and D_2. The gate operates as an OR gate for outputs of M_1 and M_2, one of the outputs is released if the other is erroneous and the identical output, if both are error free. In other words, it acts as a parallel redundant system with the minimum of one module working. If this system is designated as DR(2) then the reliability $R^*[DR(2)]$ is given by

$$R^*[DR(2)] = 2RR_D - R^2R_D^2 \text{ -- (4.1)}$$

M1,M2 – two identical units, D1,D2 – error detectors, V – validation gate

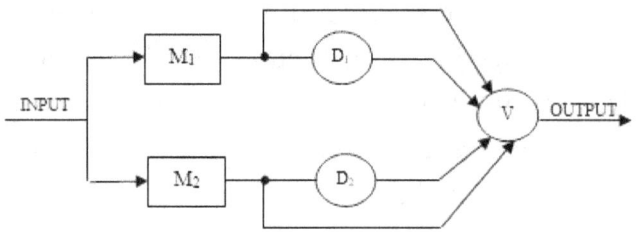

Fig.4.1 Two unit redundant system with error-detection

where 'R_D' is the reliability of the detector which is again the product of the reliability of the error detection algorithm (R_e) and the reliability of the hardware part (R_H), 'R' is the reliability of the processor module. This dual redundant system with error detection can be compared to a triple modular system with 2 out of 3 voter (minimum of two modules have to be working at any time) and the reliability of such a TMR system is denoted as $R^*(TMR)$ which is calculated as follows,

$$R*(TMR) = 1 - \sum_{i=0}^{1}\binom{3}{i}R^i(1-R)^{3-i} = 3R^2 - 2R^3 \text{ --------------------------------- (4.2)}$$

Now $R^*[DR(2)] - R^*(TMR)$

$\quad = 2R - R^2 - 3R^2 + 2R^3$ (assuming $R_D = 1$ for the present)

$\quad = 2R - R^2 - 3R^2(1-R) - R^3$

$\quad = 2R(1-R)^2$ -- (4.3)

$2R(1-R)^2$ is always positive and therefore $R^*[DR(2)] > R^*(TMR)$ for any R $(0 < R < 1)$ Even if $R_D \neq 1$ is assumed, critical value for R_D at which the above inequality is true can be calculated. The necessary and sufficient condition for

$R^*[DR(2)] > R^*(TMR)$ is

$$\frac{1 - (2R+1)^{1/2}(1-R)}{R} \leq R_D \leq 1. \text{ --- (4.4)}$$

Similarly it can be proven that the reliability of N modular detector system is greater than that of a M out of N modular redundant system i.e.

$$R^*[DR(N)] \geq R^*(NMR) \text{ -- (4.5)}$$

with the value of R_D varying between 1 and the critical limit.

4.2.1 With the addition of error detectors, not only, is the reliability increased as shown, but also most faults are detected immediately as they happen, facilitating rectification and preventing the system reaching an unsafe state. This enhances safety by reducing the probability of unwanted or unacceptable failures.

4.3 A 2 out of 2 system

The reliability of a 2 out of 2 system with two processor modules and a comparator (which is either a hardware or software detector) is reduced because of the condition that the system will work with both the modules being active and if the outputs are incongruent, the system is shut down. The reliability applying the M out of N redundancy formula is given by

$$R^*[2|2] = R \text{ -- (4.6)}$$

as compared to a plain redundant system where $(R^*[dr] = R^*(\text{duplicate redundancy}))$

$$R^*[dr] = 2R - R^2 \text{ -- (4.7)}$$

If R = 0.9 then $R^*[2|2] = 0.9$ whereas

$R^*[dr] = 0.99$

It is however to be noted though the reliability is reduced, safety is improved, as a failure in one module is taken as a failure of the whole system and it is repaired before a seccond failure takes place without giving rise to common mode failure and an unwanted response.

4.3.1 Cold and hot stand by redundancies

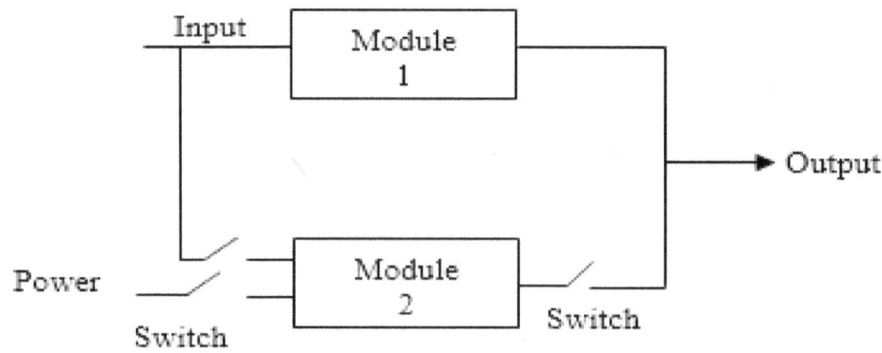

Fig. 4.2 cold standby redundancy

In the cold standby redundancy the module 2 is switched on only when module 1 fails and till such time it is dormant without connection to the power. Assuming the reliability of the switches as unity, the reliability R of the combination is given as

$$R(t) = e^{-\lambda_1 t} + \lambda_2 t e^{-\lambda_2 t} \text{ where } \lambda = \text{constant failure rate} \quad \text{-- (4.8)}$$

If $\lambda_1 = \lambda_2 = 0.1$ failures per 1000 h

$$\text{Then system reliability} = e^{-\lambda_1 t} + \lambda_2 t e^{-\lambda_2 t} = 0.9048 + 0.1 \times 0.9048 = 9953.$$

This is higher than that of the hot standby or active redundant system for which R = 0.9909, since the cold standby system works for a shorter time i.e.is at risk for a shorter time. However hot standby is preferred as the switching time of redundant output is negligible. Also the dormant system in cold standby may also reveal an unexpected failure.

4.4 Monitoring of safety critical software

As the vital outputs of an electronic interlocking system are generated by the inherent software in the system, the system safety is dependent on the software and how 'safe' it is. "Software is considered safety-critical if it controls or monitors hazardous or safety critical hardware or software" as per NASA technical standard [2]. A hazardous situation arises when there is a chance of a mishap occurring which leads to injury, death, destruction or loss of vital equipment or damage to the environment. The cause for a hazard to happen is not just a defect in hardware or software but also an unexpected input or event which results in a hazard. For continuously controlled system an operator error can also cause a hazard. To prevent the occurrence of hazardous situations, the software has to be monitored primarily in addition to the monitoring of hardware. Software is the first line of defence as it can respond quickly to potential problems and can provide more functionality than equivalent hardware. The software can detect on error much ahead displaying a message to the operator and the hardware control can be used as a backup.

4.4.1 Monitoring of software through diagnostic programs, self-testing programs and other checks and precautions to prevent software entering an unwanted or unsafe state has now become common and electronic interlocking system manufacturers are incorporating these in all their systems irrespective of whether the system consists of a single processor, duplex processors or multiple processors with majority voting systems to enhance safety and revert the system quickly to a known safe condition upon detection of unsafe conditions.

4.4.2 There are many practices developed over the years which if incorporated into the design or implementation, will increase the safety of the software. These have been collected from various sources and listed in NASA's guidebook at [2].

(i) **CPU self-test**:- The CPU could be damaged due to electro-magnetic interference, electrical discharge, shock etc., To check this at the booting time itself, CPU self test routine is run and if it fails, the software can be brought to a safe state.

(ii) **Guarding against illegal jumps in memory**:- In memory (RAM usually) the program may jump accidentally into a halt or illegal instruction, in which case, this should lead to a trap vector (one of the interrupt vectors) which in turn will point to a program which puts the system in a safe state.

(iii) **ROM tests**:- Before executing the software in ROM (also EEPROM, flash disk) it is necessary to verify its integrity and this is done at power-up, after the CPU self test and before the software is loaded for execution.

(iv) **Watchdog timers**:- A watchdog timer resets or reboots the CPU if it is not 'tickled' within a set period of time. Interrupt is not to be used to tickle the watchdog.

(v) **Stack checks**:- Checking the stack, guards against stack overflow or corruption. A stack monitor function can be used to watch the amount of available stack space and if it is limited, an error processing routine can be called.

4.4.3 Some of the problems faced by real time developers applicable to safety and reliability as per D.B. Stewart are as follows:-

(i) *Delays implemented as empty loops*:- This can create problems (and timing difficulties) if the code is run on faster or slower machines

(ii) *One big loop*:- A single large loop forces all parts of the software to operate at the same rate. This is usually not desirable.

(iii) *Error detection and handling are an afterthought and implemented through trial and error*:- The error detection and handling mechanisms should be designed from the start. These should be put at critical locations where data needs to be right or areas where the software or hardware is especially vulnerable to bad input or output.

(iv) *Indiscriminate use of interrupts*:- Use of interrupts can cause priority inversion in real time systems if not implemented carefully. Interrupts should be avoided wherever possible.

As per B. Wood, one of the practices for software risk management is

* *Use read backs to check values*:- When a value is written to memory, the display, or hardware, another function should read it back and verify that the correct value was written.

4.4.4 In addition to the above, the following can be considered to enhance safety.

(i) *Reduce complexity*:- Complex components can be reduced if possible. McCabe's cyclomatic complexity can be calculated and if it is over ten, the routine can be simplified.

(ii) *Prohibit program patches*:- During development, patching a program should be avoided. Changes in the code should be made and recompiled

(iii) *Create a list of possible hardware failures that may impact the software*:- The software must respond properly to these failures. Having a list makes explicit what the software can and cannot handle.

4.4.5 Some of the following programming suggestions are from SSP50038, computer based control system safety requirements for the International Space Station Program.

(i) Provide for an orderly system shut down (or other acceptable response) upon the detection of unsafe conditions. The system can revert to a known, predictable and safe condition upon detection of an anomaly.

(ii) Hazardous sequences should not be initiated by a single key board entry. This means that at least two keys should be pressed for a safety critical function.

(iii) Prevent inadvertent entry into a critical routine:- Such entry if it occurs, should be reverted to a known safe state.

(iv) Do not use a 'stop' or 'halt' instruction:- The CPU should be always executing, whether idling or actively processing.

(v) When possible, put safety critical operational software instructions in nonvolatile ROM.

(vi) Verify critical commands prior to transmission and upon reception. This double check is very useful in safety critical applications

(vii) Decision logic using data from hardware or other software components should not be based on values of all ones or all zeros. Use specific binary patterns to reduce the likelihood of malfunctioning hardware/software satisfying the decision logic.

(viii) Safety critical components should have only one entry and one exit point.

(ix) Always initialize the software into a known safe state. This implies making sure all variables are set to an initial value and not the previous value prior to reset.

4.4.6 The following requirements for safety critical software have been selected from the reference at [9]

(i) Fault detection programs shall be designed to detect potential safety-critical failures prior to the execution of the related safety critical function. Fault isolation programs shall be designed to isolate the fault to the lowest level practical and provide this information to the operator or maintainer.

(ii) Data transfer messages shall be of predetermined format and content. Each transfer shall contain a word or character string indicating the message length (if variable), the type of data and content of the message. As a minimum, parity checks and checksums shall be used for verification of correct data transfer. CRCs (Cyclic Redundancy Checks) shall be used where practical. No information from data transfer messages shall be used prior to verification of correct data transfer.

(iii) External functions requiring two or more safety critical signals from the software shall not receive all of the necessary signals from a single input/output register or buffer.

(iv) Alerts shall be designed such that routine alerts are readily distinguished from safety critical alerts. The operator shall not be able to clear a safety critical alert without taking corrective action or performing subsequent actions required to complete the ongoing operation.

4.5 Economy of single processor system

Monitoring of software as detailed in the earlier paragraphs has become imperative to improve the safety of the microprocessor controlled systems and the reliability of a 2 out of 2 system with a comparator is not enhanced as shown by Equation 4.6. Also there is a likelihood of common mode failures happening simultaneously in such a system. The provision of a redundant processor module and a processor module for the comparator can be considered infructuous as there is no commensurate benefit derived. These can therefore be dispensed with and a single processor system with the self-testing and diagnostic routines will definitely be more economical and sufficient.

4.5.1 Similarly the reliability improvement in a Triple Modular Redundancy (with majority voting) is not substantial. $R^*(TMR) = 3R^2 - 2R^3$ and if R = 0.9, $R^*(TMR) = 0.972$ which is not very high and is achieved because of a fault tolerance when one module fails and the system works as a 2 out of 2 system. Again there is the problem of common mode failures. The provision of two redundant processor modules and a processor module for majority voting will of course be costly considering the benefits accrued. Hence a single processor system with the requisite self checking routines proves to be economical.

4.6 Single processor systems

Single processor electronic interlocking systems were developed initially in the U.S.A by two signalling firms namely Union Switch & Signal (now M/s. Ansaldo) and General Railway Signalling Co (now M/s. Alstom). The systems developed by U.S. & S. originally Microlok I, later Microlok II have been installed extensively on the Indian Railways starting from 1994. The details of Microlok II have been described in Chapter 3.

4.6.1 The system developed by GRS (General Railway Signal) Co. was also installed on American Railways from the early 1980s onwards and later on the World Railways, such as, Indonesian Railways etc. A similar system has been installed on Delhi Metro Railway. A notable characteristic of this system is that the decision logic is based on specific binary patterns and not all ones or zeroes as per the suggestion (vii) in paragraph 4.4.5 above, based on SSP50038 and the requirements (ii) of paragraph 4.4.6 from the reference [9]. Polynomial codes have been used extensively in this system to enhance security or safety. The Vital Processor Interlocking (VPI) in this system is described in the next paragraph with reference to [7] and other open sources.

4.6.2 Vital Processor Interlocking of GRS Co.

The processes involved in the VPI are illustrated in the figure 4.3. The vital inputs are read by the input circuits and coded as a testword every 1 sec and circulated through a vital input check circuit serially. The input check circuit is energized in the field through a positive supply to the transistor circuit buffer and the inverse of the test word is sent back to the CPU. This happens similarly in the second diverse channel as two channels are employed for verification. The non-vital inputs are also read but not passed through the check circuit. The check circuit for the vital inputs checks the absence of high levels of induced A.C. on the field conductors. The input check words received are taken as the code words and are processed as per the interlocking logic stored in EPROM typical for the station. Here, logic operations are evaluated through specific algorithms based on polynomial bit weightage and division, so that a higher degree of security is achieved compared to a simple logic scheme (AND, OR, ADD etc). This is done for both the diverse channels, once every second.

4.6.3 The output data word coming out after the boolean interlocking logic evaluation is sent as a testword to the 'absence of current detector' (AOCD) to check whether there is no current in the output circuit or below a limit. The circuit is similar to a transformer circuit with three windings and 'data in' is transmitted to 'data out' from one winding to another without change, only if the current in the third winding is low. The testwords coming out of the AOCD circuit serially as codewords are now again compared with the output codewords of the logic evaluation. This is done every 50 ms i.e. 20 times in an operational cycle of 1sec. If at any time, there is incongruity, the output to the vital cutoff relay is cut off shutting down the system.

4.6.4 The vital processor is a 16 bit processor with a 32 bit shift register attached, which does all the polynomial division required at each stage. Out of the 32 bits, the lower 14 bits carry the information and the upper 18 bits are used for check field. For checking vital checkwords, 18th order polynomial is used as divisor and for memory check, a 32 bit polynomial is used. To process a 32 bit word, the 16 bit processor employs two stage processing.

4.6.5 Vital relay driver

The vital relay driver board consists of an 8 bit processor which checks the validity of the checkwords and produces an output voltage capable of operating a Q type relay. The main processing system creates a set of data called "main checkwords" once every second and a set of recheck checkwords every 50 ms (20 in all in 1 sec interval) based on the status of all the system's outputs. If any of these checkwords are not correct, the power to vital cutoff relay is cutoff. When all checkwords are correct at the end of 1 sec cycle, the processor releases a 10 KHz square wave modulated at 500 Hz.

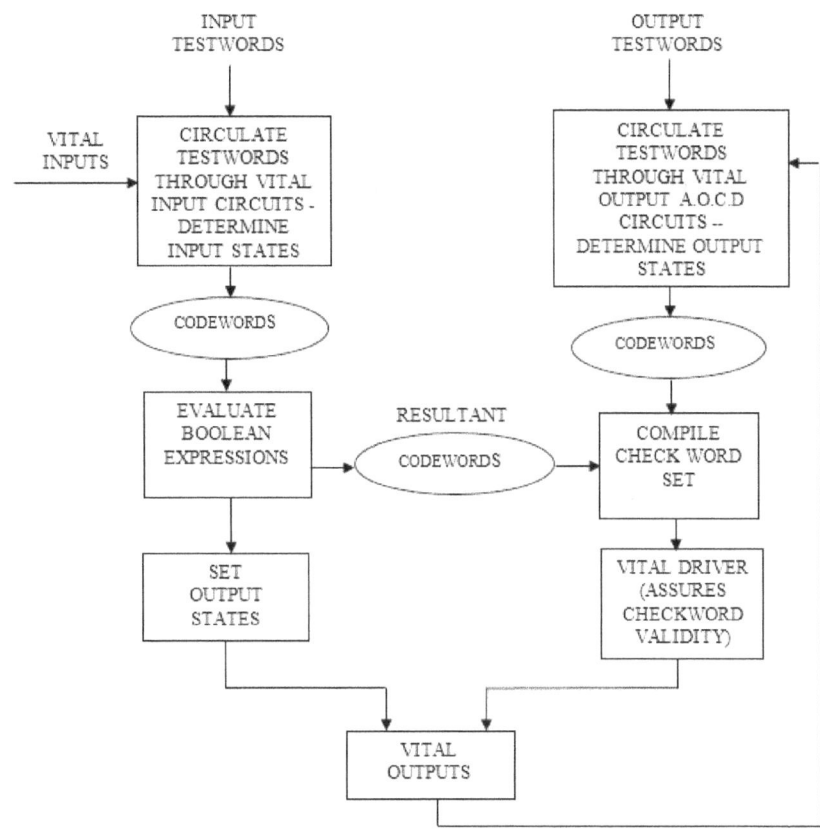

Figure 4.3 Vital processes in the VPI (GRSCo.)

After passing through tuned circuits, the wave is rectified and filtered to produce the required D.C. Voltage. This process takes about 50 msec. The vital relay driver thus uses Numerically Integrated Safety Assurance Logic (NISAL) and cannot produce the correct output without the checkwords being absolutely correct.

4.7 A typical error-detection method

Over the years, all manufacturers of electronic interlocking systems have incorporated some sort of error detection methods and diagnostic routines to enhance the security or safety of the software and thereby the system, irrespective of the hardware redundancies they have employed and the cost factor involved. An error-detection method proposed by the author and read at the International Conference on "Railway Safety, Control and Automation towards the 21st Century" organized by the Institute of Railway Signal Engineers (London), 25–27 September 1984, referred to at [4] of chapter 3 has been summarized in Appendix A.

4.7.1 The interlocking requirements or say a line in the interlocking table can be equated to an 'AND' gate or a series of 'AND' gates with sequential inputs. An error in an 'AND' gate can be detected by the presence of synchronous pulses on all the inputs and the output and under certain conditions the presence of a pulse on one of the inputs and the output. This is also confirmed by the concept of multiple boolean difference for 'AND' gate.

$$\frac{d^n F(x_i)}{dx_1 \, dx_2 \ldots dx_n} = F(x_1, x_2, x_3 \ldots x_n) + F(\overline{x}_1, \overline{x}_2, \overline{x}_3 \ldots \overline{x}_n) \tag{4.9}$$

where $F(x_i)$ = output of 'AND' gate with inputs $x_1, x_2 \ldots x_n$. The multiple boolean difference is equal to unity (requirement for testability) only if the inputs are all simultaneously either '1' or '0'. In a microprocessor controlled interlocking, the interlocking is in the form of 'firmware' or software but the ability to read a pulse at one of the inputs of a port of a peripheral chip and to generate an identical pulse at one of the outputs of a port of another peripheral chip is considered a healthy condition of the system, as the inputs in a signalling system are binary and the faults which affect the system are either 'stuck-at-0' or 'stuck-at-1' faults. The absence of a pulse at the output can be considered an erroneous condition of the system. As the object of the microprocessor interlocking system is to have a failsafe output, the system shall be either totally self-checking i.e. all the faults in the system shall be detectable or alternatively partially self-checking and partially fault-secure i.e. the fault which cannot be detected should not affect the output at all, so that the output is always predictable. An experiment had been conducted with this method and the results including a quantitative assessment have been detailed in Appendix A.

4.8 Quantification of hazard probability

As per the Department of Defense (USA) Standard MIL-STD-882E[1], the probability of occurrence for the improbable category – level E is less than 1 in 10^6. If probability per hour is to be taken over a 100, 000 hour system life it is less than 10^{-11} as suggested in [10]. As per the FAA (Federal Aviation Administration) USA document no. AC25.1309–1A[8] under extremely improbable failure conditions, the failure probability is $<10^{-9}$. It is taken as 10^9 working hours.

4.8.1 As per CENELEC (French acronym for European Committee for Electrotechnical Standardization) Standard EN 50129, for SIL (System Integrity level) 4, the Tolerable Hazard Rate (THR) per hour and per function is between 1 in 10^9 and 1 in 10^8. The Indian Railways have prescribed SIL4 for all electronic interlocking systems as per RDSO/SPN/192/2005.

4.8.2 As per the guidelines of the Department of Defense and other organizations in U.S.A the quantitative figures of failure probabilities under various conditions cannot be taken as guaranteed values and can be considered only for relative evaluation of safety just as the figures for reliability which are also based on the theory of probability. The figures calculated by adopting FTA (Fault Tree Analysis) or other methods are generally arrived at based on many assumptions and cannot give a very accurate picture of safety involved. Also many factors whose probability is difficult to estimate crop up during investigation of a mishap. Both qualitative and quantitative evaluations of safety are therefore required for obtaining optimum safety.

4.8.3 *Evaluation of Hazard rate for the experiment described in Appendix A.*

The fault tree for accident probability for the experimental system at model station 'X' is given in Fig. A.4. For the evaluation of hazard rate only the electronic control portion of the system is taken into account, i.e. the events, A, B, C leading to undetected wrong side failure are considered and not the whole system. Let the probability of intermediate event leading to wrong side failure be termed P_{WSF}. Now $P_{WSF} = P_A + P_B + P_C$. It is seen that the occurrence of the events B & C is negligible as per reasoning given in [4] of chapter 3 and $P_{WSF} \simeq P_A$. The failure rate (F) of the electronic control portion is evaluated to be 0.5724×10^{-6} per hour taking the figures for each hardware component from MIL-HDBK-217F referred at [6]. The lifetime of the equipment is taken as 20 yrs (L) and the number of failures N will be

$$N = F \times L \times 365 \times 24 = 0.5724 \times 10^{-6} \times 20 \times 365 \times 24 = 0.1 \quad \text{------------------------------- (4.10)}$$

In the experiment conducted for error detection, 63% of the artificial faults could be detected and 37% faults though could not be detected, did not affect the system. Assuming that the undetected faults have the potential to cause wrong side failures, P_1 being the proportion of such failures, the number of wrong side failures during the lifetime is P_1N. The mean time between wrong side failures W in hours is given by

$$W = \frac{L \times 365 \times 24}{P_1 N} = \frac{20 \times 365 \times 24}{0.37 \times 0.1} = 4735135$$

The probability of undetected failure of diagnostic program P_A is given by

$$P_A = \frac{P_1 N}{W} = \frac{0.37 \times 0.1}{4735135} = 0.78 \times 10^{-8} \quad \text{--- (4.11)}$$

Here $P_A \simeq P_{WSF}$, therefore $P_{WSF} = 0.78 \times 10^{-8}$ (per hour) which is above the figure of 1 in 10^9 per working hour as per FAA standards. To reduce this probability, a method suggested in [5] of chapter 3, such as duplicating the input channels or duplication of other components and/or other methods can be employed. However referring to the standard EN 50129 (in the next paragraph), it gives a better picture.

4.8.4 Hazard rate as per EN 50129

The failure rate (F) of the electronic control system as already evaluated is $F = 0.5724 \times 10^{-6}$ (per hour). It is seen that 37% of failures cannot be detected. The potential Hazardous Failure Rate (HFR) = $0.37 \times 0.5724 \times 10^{-6} = 0.211 \times 10^{-6}$. In the experiment, it is assumed that the fault can be detected well within 1 sec and negation also can be possible with the large period of 1sec, the Safe Down Time (SDT) can be taken as 1sec. Hence the Tolerable Hazard Rate (THR) is given by

$$THR = HFR \times SDT = 0.211 \times 10^{-6} \times 1/3600 = 5.86 \times 10^{-11} \quad \text{------------------------------ (4.12)}$$

This figure of THR is well below the value prescribed for SIL 4 of EN 50129. If the error detection and negation work as envisaged, there is no need to alter the system.

4.8.5 The quantitative measure for catastrophic or improbable occurrence was not clearly spelt out in the early 80's when the paper at [5] of chapter 3 was published, even as per UIC (International Union of Railways) and Aviation Authority, the quantitative recommendation for hazard rate and improbable occurrence wherein loss to life and property takes place came out in the late 90's and as already stated, the figures mentioned should not be taken as a guarantee of safety but as a guidance for evaluating the performance.

4.8.6 In the calculations for hazard rate in paragraph 4.8.4, it is assumed that the probability of software error is almost zero. As per Ms Lutz and Ms Leveson, software errors are mostly due to discrepancies in requirements specification and in designing software interface with the rest of the system and not coding errors. With exhaustive functional testing, white box and black box testing, coding errors are virtually eliminated. Also with verification and validation by an independent authority, the errors in software requirements and their interpretation can be eliminated. The Indian Railways have now Electronic Interlocking systems which are about two decades old and no hazardous failure of hardware or software of the electronic equipment has been reported. Mostly failures of external equipment which are electro-mechanical have been reported.

4.9 Study of failures of some systems

E.I.s of U.S&S make have been installed at eleven stations on the east coast of India on the South Central Railway starting from 1994. A study was conducted between 2003 and 2009 and the total failures (of the whole system) amounted to 125 out of which transient failures were 46 and card failures were 42. These can be considered only random failures as per SIL 4. None of these are considered to be unsafe. This type of E.I was extended to 46 stations gradually on this railway and a random study of the failures in March 2015 reveals that the failures were only 2 in the month. The failures per station per month in the earlier survey gives a figure of 0.1578 which has come down to 0.0435 in March 2015. This shows that the period of 'infant mortality' in the 'bath tub' curve has passed and the present phase is in the flat portion. However failures due to lightning damage in some stations in the eastern region could not be arrested in spite of providing robust earthing and surge protection. Efforts are still being made to find a solution to this problem.

4.9.1 Comments about systemic faults

As per paragraph A.3 of EN 50129, non-quantifiable systematic, (more appropriately systemic) faults are caused by human errors relating to 1) specification 2) design 3) manufacture 4) installation 5) operation 6) maintenance and 7) modification. These are eliminated by quality and safety management at various stages of the system life cycle (from EN 50126) shown in Fig. 4.4 modified suitably. Even if a few such failures appear in the 'infant mortality' stage of the installation, these can be rectified permanently.

4.10 Quality and Safety Management in the Indian context

When an indigenous manufacturer offers the EI for the first time, the equipment is visually inspected by RDSO and if it meets the physical and general requirements of hardware/software including the architecture etc., based on preliminary observations, the equipment is placed at a station yard with the outputs connected in parallel with the outputs of the existing relay based interlocking system. The application software is coded to correspond to the relay logic already existing. The correspondence of the outputs is monitored by a logger for at least 180 days and for success, the outputs should be identical. Any discrepancy is studied and it should be ensured that it is not due to the malfunction of the experimental system. If this experiment is successful, the outputs of the experimental system are connected in series with the outputs of the relay system and the output relays will be energised only if both the outputs are logically '1'. The experiment is continued for 180 days and if successful, the experimental system is connected on a standalone basis for a period at the discretion of the RDSO. If the system passes the three stages, it should overall comply with the Indian Standards RDSO/SPN/192/2005 for 'Electronic Interlocking' and RDSO/SPN/144/2014 for 'Safety and Reliability requirement of electronic signalling equipment'.

4.10.1 The two Indian Standards (RDSO/SPN/144 and 192) have been drafted based on the relevant standards of ISO (International Standards Organization), IEC (International Electro-technical Commission) and CENELEC (European Committee for Electro-technical Standardization). The steps taken to mitigate the systemic faults by quality and safety management in the Indian context are described briefly. The concept, system definition and system requirements are finalised based on the interlocking plan, selection table or control table and panel diagram of the station, which are to be approved by the competent authority after thorough checking at least at two levels. The other system requirements, such as, its suitability for working in 25 KV A.C. traction areas, under temperature conditions of-40°C to 85°C, user friendly graphic based design tool etc. are verified by R.D.S.O. The compliance to SIL4 of CENELEC standard has to be certified by the local firm and if the vendor is a foreign supplier, he has to submit a certificate that the system is equivalent to SIL 4. Detailed requirements have been listed for protection against electromagnetic and electrostatic interference in Specn. 144. Compliance of the system software and application software and their storage are verified by RDSO, also the self-check procedures as laid down in Specn.144. Under 'risk analysis', the vendor shall furnish the details about the methods adopted for fail-safety e.g. use of a watch dog timer, automatic shutdown etc., proof of safety in the form of process adopted for safety analysis and result thereof. Full documentation of software engineering followed during development has to be supplied for verification. Quality Assurance Program along with a certificate from in-house Quality Assurance (QA) group or an Independent Safety Auditor (ISA) is required for safety and risk Analysis. A foreign supplier has to submit a lot of information for cross-acceptance as per clause 10(j) of Specn. 192. The Design and Manufacture stages should be based on the requirements given in specifications 144 and 192 regarding the housing of the cards, wiring etc., power supply requirements, also the plant and machinery at the indigenous manufacturer have to be approved by the RDSO. For a foreign supplier, a type approval certificate for the plant and machinery and testing equipments used abroad has to be produced.

4.10.2 Tests

Card level functional tests on all the cards and fail-safety tests on one card of each type are conducted. After successful integration of the modules, system level functional tests and fail-safety tests are conducted. Computerised testing for a minimum of two hundred thousand permutations and combinations is to be done. Environmental and climatic tests are conducted in an approved laboratory. System diagnostics test is conducted by automatic test procedure through a PC.

Checksum of software is verified and the application program is verified for congruence with the selection table by the user of the equipment.

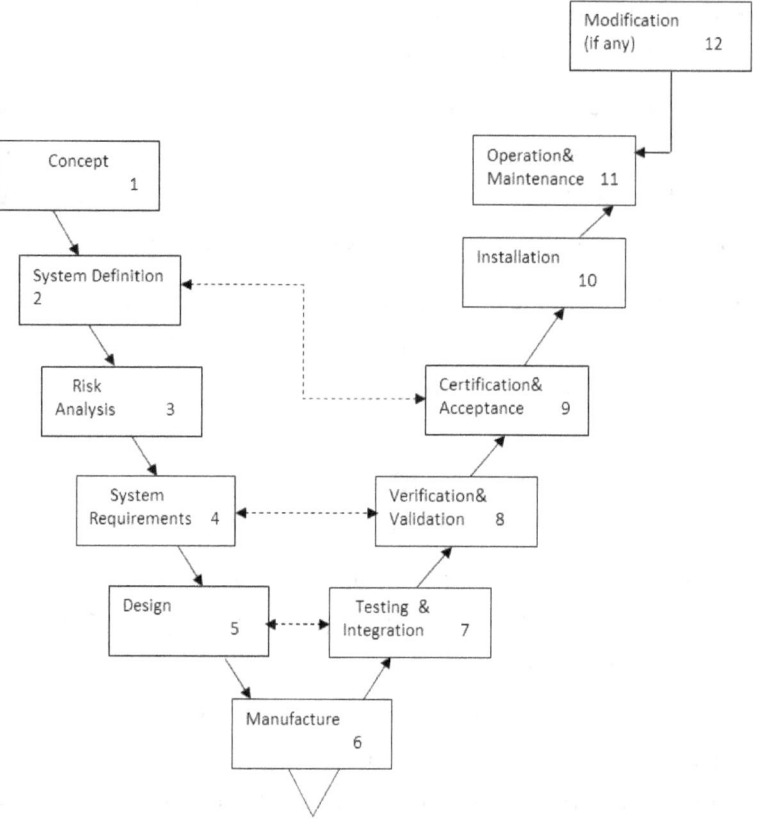

Fig. 4.4 'V' Development Life Cycle Model

(Note: The dotted lines connecting the two halves represent verification, testing and consolidation of the processes on the left side)

4.10.3 Verification and validation

In addition to testing carried out by RDSO, verification and validation of hardware is done by an approved independent agency in India for the indigenous equipment, which includes fail-safe testing, environmental and electro-magnetic interference testing and other tests, such as, verification of VHDL model of CPU and simulation tests etc. The software is also verified and validated for compliance to CENELEC 50128 (SIL 4) by an approved external agency to the satisfaction of RDSO. Regarding the EI of U.S & S., the hardware and software was verified and validated by Battelle Agency of Columbus Ohio and the firm has itself developed a safety validation tool called D_RAMP. Similarly other foreign suppliers have submitted certificates from competent authorities regarding the completion of verification and validation of hardware and software to the satisfaction of the RDSO.

4.10.4 Certification and acceptance

Acceptance tests are carried out before certification. These tests include 1) visual inspection 2) insulation resistance tests 3) card level functional test on all the cards 4) system level functional tests 5) System diagnostics test and 6) verification of application software based on selection table which is normally done by the user railway. Visual inspection includes physical dimensional checks, housing and shielding of components etc. Insulation resistance test is conducted with 500V D.C. and the insulation resistance shall not be less than 10 mega ohms. Other tests have been described at the detailed testing stage.

4.11 Installation, operation, maintenance and modifications

The manufacturer will have to supply installation and maintenance manual with pre-commissioning checklist and diagnostic aids including troubleshooting charts. Step by step actions to rectify the faults in case of failures should be detailed so that the maintainer can easily replace the defective cards or components. Details of the system circuits and layout of PCBs etc. in addition to the details of software flow chart etc. shall be given to the user. Maintenance Terminal (MT) consisting of a standard PC with printer shall be provided for display of the current status of the yard, storage and display of recorded events and provision to transfer information to an external data logger. Modification if any, both in hardware and software should have the approval of competent authority and should be carried out only after required testing and validation.

4.11.1 No systemic faults due to human errors in the life of the present EI systems installed on Indian Railways have been reported so far. This includes the indigenous system (Medha & Co) which has been in existence for almost seven years. The only failures reported are due to random faults which are also declining due to the systematic maintenance being carried out by staff who have gained experience. Failures can be further reduced by standardising the external wiring for the system and providing rigorous training to the skilled staff in the modern technologies. Now, EI is proposed for large yards also where the routes are more than 200.

4.12 E.I. for large yards

The R.D.S.O has issued a specification no RDSO/SPN/203/2011 for installation of E.I. at large yards having more than 200 routes. The specification no. RDSO/SPN/192/2005 is applicable to EIs upto 200 routes only. The EI for large yards is designed to have more than one EI module with communication of safety information complying with EN50159 standard. Each EI module is limited to control of upto 800 inputs or outputs. Also object controllers may be used to control some portions of the yard. Some modifications in the architecture have been incorporated to increase the reliability. Though partial VDU screens can be provided at multiple cabins for visualisation of the relevant portion of the yard, the operation of the yard is to be from a central cabin with provision of a large VDU screen to see at a glance the whole of the yard.

4.13 Future Policy

All future interlocking for yards with more than 50 routes are to be provided with E.I. to the specifications already in vogue. Even at yards with less than 50 routes, E.I.s can be provided if there is sufficient financial justification.

References

[1] Department of Defense (USA) – System Safety Standard – MIL-STD 882E – 11th May 2012.

[2] NASA-GB-8719.13 "NASA Software Safety Guidebook" issued by National Aeronautics and Space Administration as a Technical Standard 31st March 2004.

[3] Neil Storey "Safety-Critical Computer Systems" Addison-Wesley Longman-1996.

[4] D.T. O'Connor, David Newton and Richard Bromley "Practical Reliability Engineering" IV edition John Wiley & Sons 2002.

[5] Ramamoorthy C.V. & Yih-Wu Han-Sept 1975 "Reliability analysis of systems with concurrent error detection" Trans. IEEE C-24, pp 868–878.

[6] "Reliability Prediction of electronic equipment" – Military Handbook Department of Defense (USA) – MIL – HDBK-217F –2nd Dec. 1991.

[7] D.B. Rutherford – "Fail-Safe Microprocessor Interlocking – an application of numerically integrated safety assurance logic" – Proceedings of International Conference-"Railway Safety, Control and Automation towards 21st Century" – London 25–27 Sept. 1984, pp. 72–76.

[8] "Safety of High Speed Ground Transportation Systems" – Federal Rail Road Administration – US Department of Transportation – Sept. 1995.

[9] "Software System Safety Handbook" issued by the Joint Services, U.S. Department of Defense Dec. 1999.

[10] William R. Dunn "Practical Design of Safety Critical Computer Systems" Reliability Press – 2002.

## CHAPTER 5

SOME METHODS OF HAZARD ANALYSIS

5.1 Failure Mode, Effects and Criticality Analysis (FMECA)

FMECA (Failure Mode, Effects and Criticality Analysis) is a widely used method for determining the critical components which are responsible for affecting the system output or causing harm to the success of a mission. Failure modes are classified in relation to the severity of their effects, so as to allocate more resources to the critical items requiring improvement.

5.1.1 An FMECA is carried out for the hardware shown in Fig A.2 as an example. The worksheet for the example is taken from [4] and slightly modified to suit the requirements. The failure mode criticality number is calculated for each component taking into account the failure rate or probability, failure mode ratio, conditional probability of failure on output or loss of function. This provides a quantitative criticality rating for the component or function.

The failure mode criticality number is given by

$$C_m = \beta \alpha \, \lambda_p t \quad \text{--} \quad (5.1)$$

where β = conditional probability of failure of output or loss of function

   α = failure mode ratio (for an item $\sum \alpha = 1$)

   λ_p = part failure or hazard rate

   t = operating or at-risk time of item

For a subsystem or an Integrated Circuit (IC) for example $\alpha = 1$ as the sum of the failure mode ratios of each small component is always equal to one. The value of β actually gives the proportional severity of the failure of the component on the output of the system but is of course subjective in assessment. The part failure rate λ_p for each component is taken from MIL-HDBK-217F under 'benign' conditions when the equipment is housed in a permanent building and protected with air-conditioning to control the dust and humidity, except outdoor equipment. $\lambda_p t$ can also be replaced by failure probability $1 - e^{-\alpha \lambda_p t}$. The failure effects of software are not taken as it is presumed that software faults are set right during testing and teething problems are attended to before commissioning, also FMECA is not amenable in application to software as faults in software are generally unpredictable.

5.1.2 An FMECA can be performed for different objectives which can be safety, mission success, availability etc. and the criticality number varies as per the objective. An item gets different criticality number if evaluated for safety and availability separately [3].

5.1.3 The table 5.1 shows the results of FMECA calculations for the hardware of Fig A.2 with the system criticality listed in decreasing order. The figures in column 7 of the table 5.1 give the probable effect of the failure of the component on the system output ranging from 0 to 1, with 1 indicating a definite effect on the output. The effects of failure of track relay and output drive relay have not been taken into account as these give fail-safe results and no unsafe result occurs. The results in column 10 give an idea about the improvement that can be envisaged at the design stage including duplication of some components if required.

5.2 Event Tree Analysis (ETA)

In Event Tree Analysis, the events that can affect a system are considered and their possible consequences are studied, if these events happen. Both normal and fault conditions are considered. In this method, a top-down approach is adopted starting from the event and listing the successive results till the final effect on the system is known [2]. While doing such analysis, unanticipated and unknown results can be brought out to improve the design of the system.

5.2.1 An example is considered, to best understand the making of the event tree. Nowadays, all microprocessor systems controlling the interlocking have various types of diagnostic programs irrespective of whether they are single processor, 2 out of 2 or 2 out of 3 systems, as briefly listed in the description of the systems installed on Indian Railways. In the example, it will be analysed as to what will happen if the supply voltage (V_{CC}) to the vital processor card comes down say from 5 V to 2 V. The event tree for this event is illustrated in Fig 5.1. The state of the system is recorded after taking the state of the diagnostic program, vital cut off relay and the maintainer's readiness to take action in case of an abnormality. Safe state is reached promptly when both diagnostic program and vital cut off relay operate as designed. There are clear unsafe states when the vital cutoff relay fails and maintainer is not attentive, also, when all the three fail. The probability of unsafe states is low and if the maintainer is attentive and if an alarm can be sounded through his mobile phone, the probability of reaching an unsafe state will be almost zero.

5.2.2 The conclusions as drawn in the event tree are subjective depending on the programs provided for diagnostics and actions to be taken in case of abnormalities. Actions to be taken to prevent unsafe states can be thought of at the design stage after analyzing the consequences as in an event tree.

5.3 Fault Tree Analysis (FTA)

Fault tree analysis is a graphical method where failure causing events are taken at the bottom and worked bottom up to the most hazardous or catastrophic event through intermediate events combining wherever necessary using logical gates. This is in contrast to the Event tree analysis in which hazardous events are selected and worked top-down to the associated failures.

5.3.1 Fault Tree Symbols

A set of elementary symbols which are commonly used in constructing a fault tree is shown in Fig 5.2. Some more symbols used in large and connected fault trees are not shown here and all these symbols follow the IEC 1025 standard.

5.3.2 Example of a fault tree

A simple door bell circuit is taken as an example for drawing the fault tree [1]. The door bell circuit is shown in Fig 5.3. The bell operates when the pushbutton is pressed and the components are in good condition i.e. the battery, switch and bell are all normal without any fault. The door bell may not operate if any one of the three items fails and this is reflected in the fault tree drawn in Fig 5.4. The bell may fail due to either (1) the clapper is broken or (2) lack of sufficient magnetomotive force in the solenoid. The battery may fail to deliver voltage due to either (1) the lack of electrolyte or (2) the positive pole is broken. The pushbutton switch may fail to make contact if either (1) the contacts are broken or (2) the contacts offer high resistance. Failure can also take place when the wire connecting the components is broken. Many more faults other than those listed in the fault tree can happen such as open/short circuit in the solenoid etc. With more experience in maintaining the bell circuit, more faults come to the fore.

5.3.3 Fault-tree for the interlocking at a model station 'X'

The fault-tree for the electronic interlocking at a model station 'X' is shown in Fig A.4. The hazardous event is an accident likely to occur when there is an undetected wrong side failure either due to non-detection of the diagnostic programs, maloperation or non-operation of diagnostic programs, at the same time an unwanted route is set during this period with the likelihood of train approaching and the operator fails to notice. From the Fig A.4 the accident probability P_{ACC} can be calculated as

$$P_{ACC} = (P_A + P_B + P_C) \times P_D \times P_E \times P_F \text{ -- (5.2)}$$

The detailed calculation given in ref [4] of chapter 3 is reproduced below.

Let the number of failures in the lifetime of the equipment of 'L' years be N. Though theoretically, the faults not detected by the error-detection method are secure and do not affect the system, it is taken that some faults remain undetected and lead to wrong side failures. Let this ratio be P_1. The ratio contributed by events D, E & F is taken as P_2. The total number of wrong side failures leading to accident in a lifetime is therefore $P_1 \times P_2 \times N$. The mean time between wrong side failures in days

$$W = \frac{L \times 365}{P_1 \times P_2 \times N} \quad \text{--- (5.3)}$$

The contribution due to event 'B' is considered to be negligible as the detector can be made fail-safe by using discrete components whose failure modes are known. If the contribution of event 'A' to wrong side failures is P_{1a} and the contribution of event 'C' is P_{1c} then $P_1 = P_{1a} + P_{1c}$. The main contributory cause in the system will be event 'A' as all the faults are not being detected by the suggested method and assuming that the system is checked every 24 hours for unrevealed failures in system by means of functional and other tests involving sophisticated instruments, the probability of event 'A' or

$$P_A = P_{1a} \times N \times \frac{1}{W \, (\text{days})} \quad \text{--- (5.4)}$$

As the run of the interlocking program is negligible compared to the response of the error-detection program, the response of the latter program will be taken into account as 't' sec. In this experiment the interlocking program takes about 3 millisecs and the error-detection program takes about 1 sec due to the delay incorporated. The probability of undetected failure during time 't' or transient failure in interval 't' is equal to

$$P_C = P_{1C} \times N \times \frac{t}{W \times 24 \times 3600} \quad \text{------------------------------- (5.5)}$$

If P_2 is the proportion of the events D, E, F causing accidents then the total accident probability is P_{ACC}

$$P_{ACC} = (P_A + P_C)P_2 \quad \text{--- (5.6)}$$

$$\text{or } P_{ACC} = P_2 \left[\frac{P_{1a} \times N}{W} + \frac{P_{1c}Nt}{W \times 8.64 \times 10^4} \right] \quad \text{----------------------- (5.7)}$$

The failure rate for the system considered comes to 16.76×10^{-6} per hour. For a system with a life of 20 years the number of failures (N) = 2.9364. If $P_1 = 0.05$ and $P_2 = 0.01$.

$$W = \frac{20 \times .365}{.05 \times .01 \times 2.9364} = 4972074.6 \text{ days} \quad \text{-------------------------- (5.8)}$$

If $P_{1a} = .04$ and $P_{1c} = 0.01$

$$P_A = \frac{.04 \times 2.9364}{4972074.6} = 2.3623 \times 10^{-8} \quad \text{-------------------------------- (5.9)}$$

$$P_C = \frac{0.01 \times 2.9364 \times 1}{4972074.6 \times 8.64 \times 10^4} = 6.8353 \times 10^{-14} \quad \text{------------------------- (5.10)}$$

then $P_{ACC} = 2.3623 \times 10^{-10}$ ---(5.11)

This value refers to the whole system including the electro-mechanical components in the outdoor yard etc. Now, confining only to the electronic control portion, as given in paragraph 4.8.3, the probability of wrong side failure $P_{WSF} = 0.78 \times 10^{-8}$. This is higher than the figure of 1×10^{-9} as per FAA of U.S.A., but taking the standard of SIL4 of EN50129 (as given in paragraph 4.8.4) the Tolerable Hazard Rate (THR) = 5.86×10^{-11} (eq.4.12) which is well below the prescribed rate. However as already stated the quantitative figures cannot be taken as a guarantee for the performance of the system and for ensuring safety both quantitative and qualitative assessments of performance have to be carried out.

5.4 Markov Analysis

Markov analysis can be applied to the fault-tree of Fig A.4. The electronic portion of the control system is taken into consideration and the probability of undetected wrong side failure is considered i.e. P_A, P_B and P_C are considered and P_D, P_E and P_F are omitted. P_B which is the probability of undetected failure during maloperation of detecting hardware/software is also omitted as it is negligible as compared to P_A and P_C. The state diagram for the Markov analysis representing the fault tree is given in Fig 5.5.

5.4.1 The state '0' is the initial state of the system and the state '4' represents the degraded state after service. The state '1' represents the state reached if the diagnostic program fails and is undetected equivalent to P_A in fault-tree. The state '2' represents the undetected failure in the period the diagnostic program is not run, equivalent to P_C. The state '3' represents the state of undetected wrong side failure. The differential equations representing the five states are given below.

$$\frac{dP_0(t)}{dt} = -(\lambda_{01} + \lambda_{02} + \lambda_{04})P_0(t) + \mu_{40}P_4(t) \quad \text{----------} \quad (5.12)$$

$$\frac{dP_1(t)}{dt} = \lambda_{01}P_0(t) - \lambda_{13}P_1(t) \quad \text{----------} \quad (5.13)$$

$$\frac{dP_2(t)}{dt} = \lambda_{02}P_0(t) - \lambda_{23}P_2(t) \quad \text{----------} \quad (5.14)$$

$$\frac{dP_3(t)}{dt} = \lambda_{13}P_1(t) + \lambda_{23}P_2(t) \quad \text{----------} \quad (5.15)$$

$$\frac{dP_4(t)}{dt} = \lambda_{04}P_0(t) - \mu_{40}P_4(t) \quad \text{----------} \quad (5.16)$$

The numerical values are

λ_{04} = electronic system failure rate = 5.724×10^{-7} per hour

μ_{40} = rate of repair = 0.33

λ_{01} = probability of reaching state '1' = 0.78×10^{-8}/hr

λ_{02} = probability of reaching state '2' = 5.866×10^{-14}

$\lambda_{13} = 1$ $\lambda_{23} = 1.$

Applying Laplace transforms and inverse Laplace transforms for solving the differential equations

$$P_0(t) = e^{-at} \text{ where } a = 0.5 \times 10^{-7} \text{ --- (5.17)}$$

$$P_1(t) = -be^{-t} + be^{-ct} \text{ where } b = 0.78 \times 10^{-8} \text{ and } c = 0.5 \times 10^{-7} \text{ ----------------- (5.18)}$$

$$P_3(t) = P_1(t) + P_2(t) \text{ -- (5.19)}$$

Taking a lifetime of 20 yrs as done in Appx. A the value of P_1 is given by

$$P_1 = 1.628 \times 10^{-22} \text{ -- (5.20)}$$

This is the probability of 'undetected wrong side failure' or

$$P_{WSF} = P_A + P_B + P_C \simeq 1.628 \times 10^{-22} \text{ ---------------------------- (5.21)}$$

where P_B and P_C are negligible compared to P_A.

5.4.2 This value of P_1 is not considered accurate as the type of assumptions required for Markov analysis do not reflect reality, for example, the failure rates and repair rates should be constant for all occurrences and the events are S-independent. This method is therefore not widely used for calculating such probabilities and the Fault Tree Analysis method is more widely used.

Table 5.1 Quantitative assessment of criticality by FMECA for hardware (Example)

1 Sl. No.	2 Hardware components	3 Failure mode	4 Failure effect	5 Abnormality	6 Failure rate λ_p	7 Conditional probability of failure of output of system (β)	8 Time (t) of use in hrs	9 Cm (system criticality) in decreasing order	10 Remarks
1.	Point motor	fails	point detector relay detects	failure of detection, effect only on some routes	13×10^{-6}	0.5	87, 600	5.69 E-1	reliability of motor and detector relays can be improved.
2.	Signal lamp (incandescent)	extinguishes	signal check relay detects	detection failure, effect only on the concerned route	51×10^{-6}	0.1	87, 600	4.46 E-1	can be improved by replacing with LEDs which is being done.
3.	Microprocessor (8085A)	halt or incorrect function	watch dog timer and self-check program detect & standby switches on	Fault not detected and standby not switched on	0.048×10^{-6}	1.0	87, 600	4.20 E-3	additional detection methods such as diverse software with complementary bits can be incorporated.
4.	Output drive circuit (solid state)	likely to get stuck	diagnostic program detects	failure of diagnostics- effect only on a particular output	0.35×10^{-6}	0.1	87, 600	3.06 E-3	reliability of the circuit and diagnostics can be improved.
5.	Interrupt interface chip (8259)	failure to run the shut down program	diagnostic program normally detects	failure of diagnostics	0.033×10^{-6}	1.0	87, 600	2.89 E-3	--
6.	RAM + I/O (8155)	bits inverted in I/O	self-check detects	self-check not able to detect, effect only on some outputs	0.05×10^{-6}	0.5	87, 600	2.19 E-3	--
7.	SRAM (2114)	bits inverted and address logic fault	memory check detects	memory check fails but other diagnostic programs may detect	0.0079×10^{-6}	0.7	87, 600	4.84 E-4	--

Supply voltage to vital processor card	Diagnostic Program	Vital Cut-off relay	Maintainer's readiness to check the error warnings.	Probability and likely State of the system	
			alert (0.8)	(0.7524) -	safe state reached
		operates (0.99)	not alert or absent (0.2)	(0.1881) -	safe state reached but maintainer not aware
	operates (0.95)		alert (0.8)	(0.0076) -	safe state reached after a delay
		fails (0.01)	not alert or absent (0.2)	(0.0019) -	unsafe state
			alert (0.8)	(0.0396) -	partially unsafe but system can be shut after seeing other warnings
		good state (0.99)	not alert or absent (0.2)	(0.0099) -	unsafe state
comes down	fails (0.05)		alert (0.8)	(0.0004) -	partially unsafe but likely to be shut with other warnings
		failed state (0.01)	not alert or absent (0.2)	(0.0001) -	unsafe state

Figure 5.1 Event tree analysis – An Example

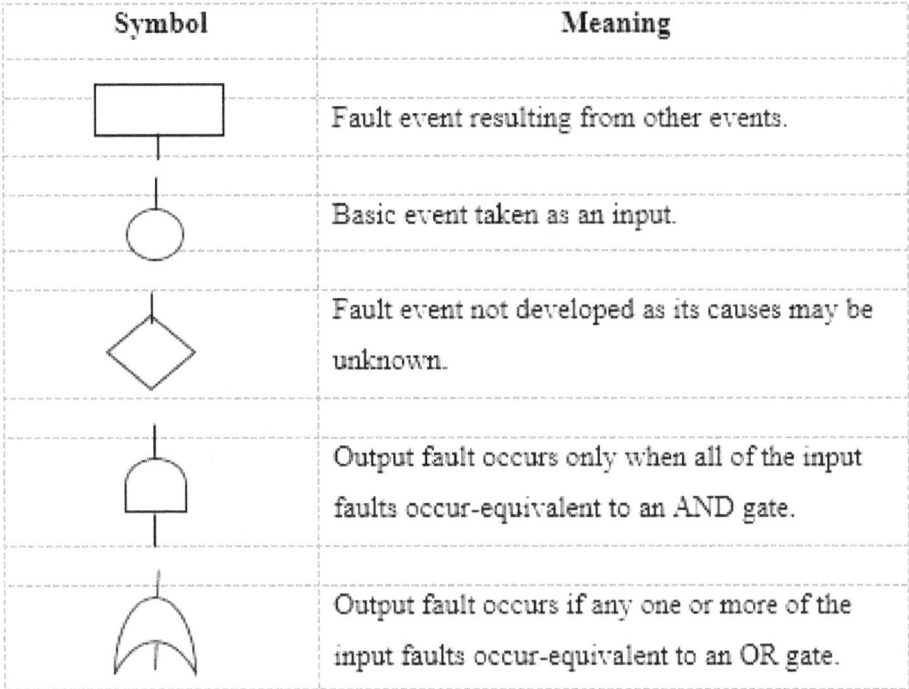

Symbol	Meaning
	Fault event resulting from other events.
	Basic event taken as an input.
	Fault event not developed as its causes may be unknown.
	Output fault occurs only when all of the input faults occur-equivalent to an AND gate.
	Output fault occurs if any one or more of the input faults occur-equivalent to an OR gate.

Fig.5.2 Fault Tree Symbols

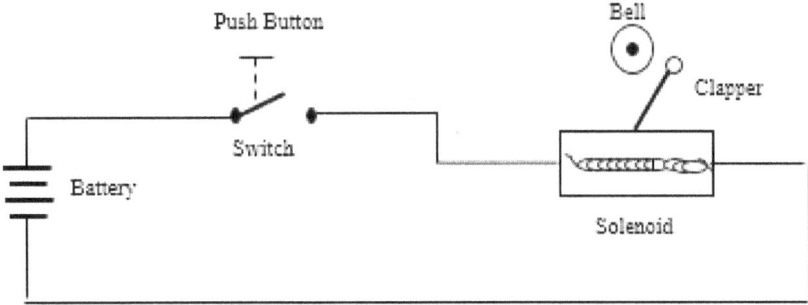

Fig.5.3 Door Bell Circuit

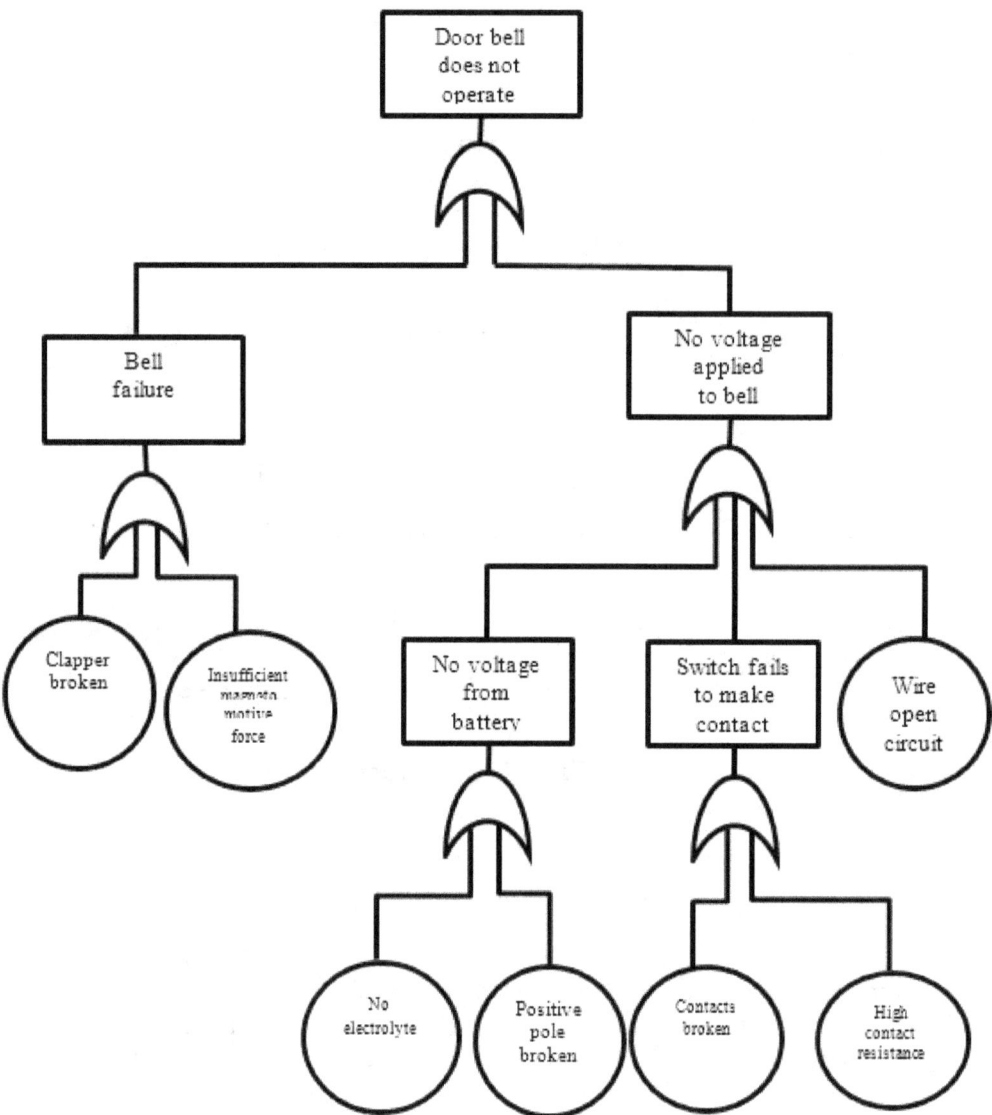

Figure 5.4 Fault Tree for door bell failure

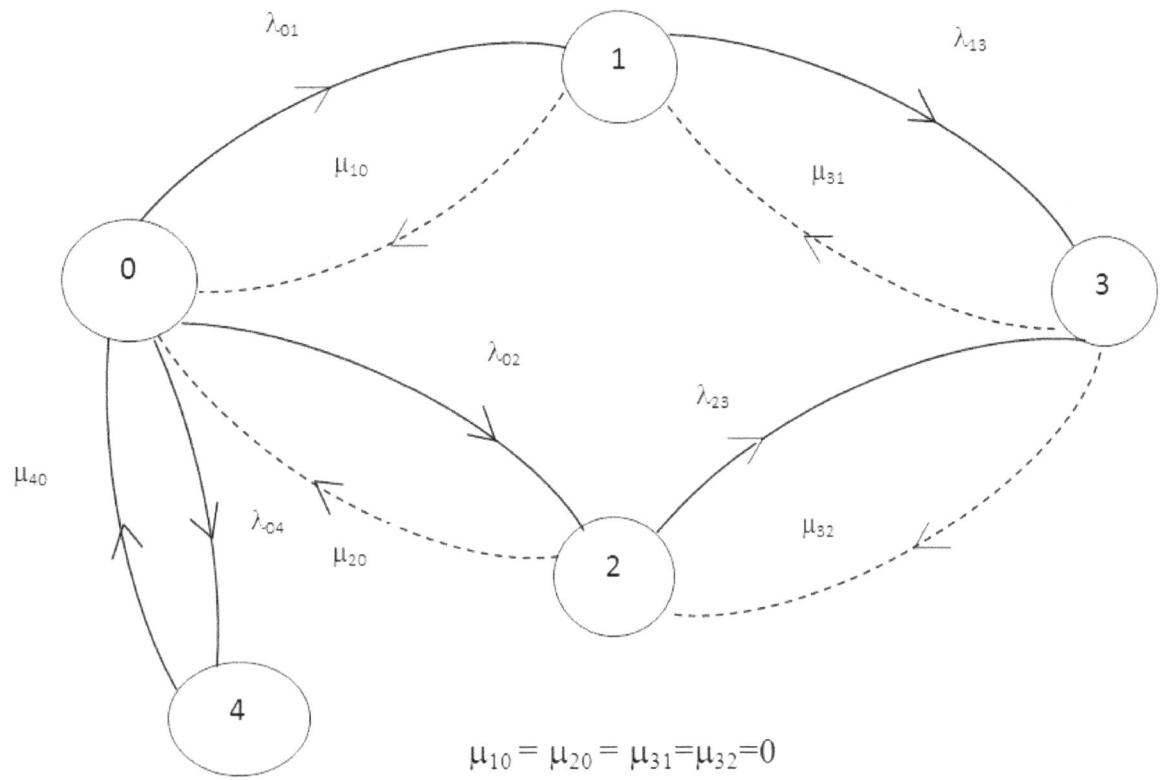

$$\mu_{10} = \mu_{20} = \mu_{31} = \mu_{32} = 0$$

Figure 5.5 State diagram

References

[1] Fault Tree Handbook issued by the U.S. Nuclear Regulatory Commission – NUREG-0492-Jan 1981.

[2] Neil Storey "Safety-critical Computer Systems" Addison-Wesley Longman-1996.

[3] Patrick D.T. O'Connor, David Newton and Richard Bromley "Practical Reliability Engineering" IV edition John Wiley & Sons – 2002.

[4] "Procedures for performing a failure mode, effects and criticality analysis"-Military Standard-Department of Defense(USA)-MIL-STD-1629A-24 Nov.1980

CHAPTER 6

HARDWARE VERIFICATION

6.1 Introduction

The dependability of a computer system is associated with the four properties namely (1) reliability (2) availability (3) security and (4) safety. While security can be achieved by various methods of cryptography, the other three properties are related to system faults which may cause unanticipated system behaviour. More emphasis should be on safety to avoid catastrophic consequences for systems such as electronic interlocking control systems.

6.1.1 Types of Faults in Hardware

As per Thomas Kropf [5] hardware faults can be segregated into (1) design faults (2) fabrication faults and (3) faults during usage. *Design* faults arise out of erroneous transformation of a design specification into the layout description which is the basis for fabrication. Hardware verification aids in eliminating these faults. *Fabrication* faults are due to layout defects during the fabrication process, which may result in incorrect behavior of the circuit. To find these faults the hardware (chips) have to be rigorously tested. *Faults during usage* arise mostly after a certain period of usage, attributed to aging, wear out and human errors during operation and maintenance. Design and fabrication faults can be eliminated by the manufacturer after verification and rigorous testing, whereas the effects of faults in usage have to be minimized by either fault-tolerance approach or self-testing and running diagnostic programs concurrently during the period of operation.

6.1.2 Importance of hardware verification

Design faults and fabrication faults if not found out at the factory premises before delivery to the customer, will prove to be very expensive as the defective lot of chips have to be recalled and the newly fabricated chips have to be delivered in replacement. On the other hand, an error found in software may be eliminated easily by recompiling the corrected program. An example is the well publicized case of the 'bug' in the floating point unit of Intel Pentium processor, where the company had to spend $475 million to replace the faulty processors. Quality control through rigorous verification at the design and fabrication stages is essential not only to eliminate hardware faults but also to reduce the time for fabrication and delivery.

6.1.3 Verification vs. Validation

Verification is the process of verifying whether the design and implementation are according to the requirements specification, while *validation* is the process of verifying whether the system meets the needs of the users in the right way. Usually both operations are performed together. Hardware simulation by means of a Hardware Description Language (HDL) or Verilog HDL can be accurate only for medium size circuits and for larger sequential circuits, higher order logic or temporal proportional logic methods are required. Model based techniques have assumed importance in verifying sequential circuits as in a microprocessor. Combinational circuits without the time factor can be verified by means of Binary Decision Diagrams (BDDs)[1].

6.2 Application of Binary Decision Diagrams (BDDs)

For example a function f = a⊕b⊕c is taken for drawing a Binary Decision Diagram (BDD). First a binary decision tree is drawn with the first variable 'a' as the root. The variable nodes at the next level will be that of 'b'. The third level will be that of 'c'. The binary decision tree for the function 'f' will be as in Fig. 6.1.

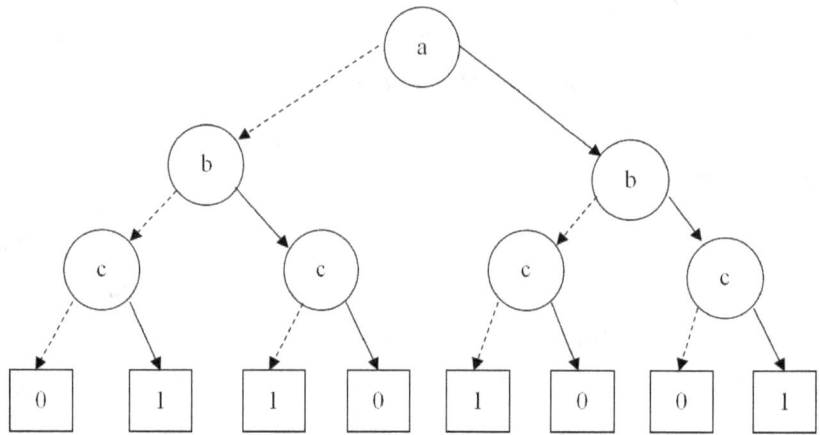

Figure 6.1 Binary decision tree for f = a⊕b⊕c

The function value is the value at the terminal, starting at root taking dashed line if the value of variable at current node is '0', taking solid line if it is '1'.

For example the value of 'f' will be taking the '1's.

$$f = \overline{a}\overline{b}c + \overline{a}b\overline{c} + a\overline{b}\overline{c} + abc \text{ or } f = \sum 1,2,4,7 \quad\text{--}\quad (6.1)$$

which agrees with the truth table for a ⊕ b ⊕ c.

A BDD is a finite DAG (Directed Acyclic Graph) with

- Unique initial node

- All non-terminals labeled with a Boolean variable

- All terminals labeled with 0 or 1

- All edges labeled with 0 (dashed edge) or 1 (solid edge)

- Each non-terminal has exactly 1 out-edge labeled 0 and 1 out-edge labeled 1.

A BDD will have only two terminal nodes i.e. representing 0 and 1. Also the BDD can be reduced to an ROBDD (Reduced Ordered Binary Decision Diagram) with the ordering of variables, in this case a, b, c based on the following.

1. *Removal of duplicate terminals*: If a BDD contains more than one terminal o-node, then redirect all edges which point to such a o-node to just one of them. Proceed in the same way with terminal nodes labeled with 1.

2. *Removal of redundant nodes*: If both outgoing edges of a node 'n' point to the same node 'm', then eliminate that node 'n', sending all its incoming edges to 'm'.

3. *Removal of duplicate non-terminals*: If two distinct nodes 'n' and 'm' in the BDD are the roots of structurally identical sub BDDs, then eliminate one of them, say 'm' and redirect all its incoming edges to the other one.

Looking at Fig. 6.1, at the 3rd level there are four 'c' nodes and two of them are identical i.e. node 1 of 'c' is identical with node 4 and node 2 is identical with node 3. Hence removing them, the reduced graph will be as follows in Fig. 6.2.

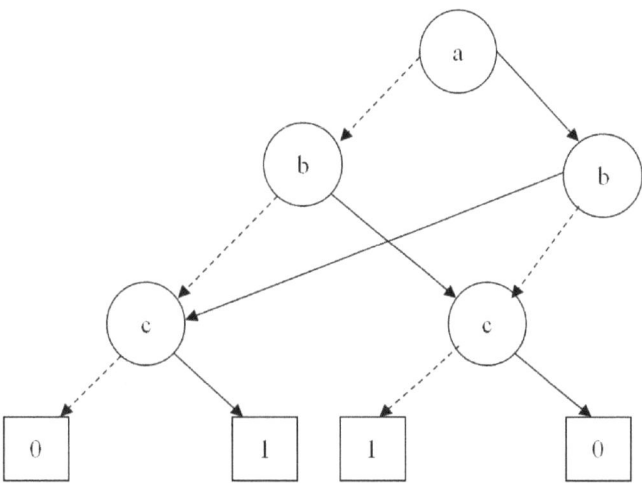

Figure 6.2 Eliminating duplicate nodes

Eliminating the duplicate terminal node '0' and '1' the ROBDD will be as in Fig. 6.3.

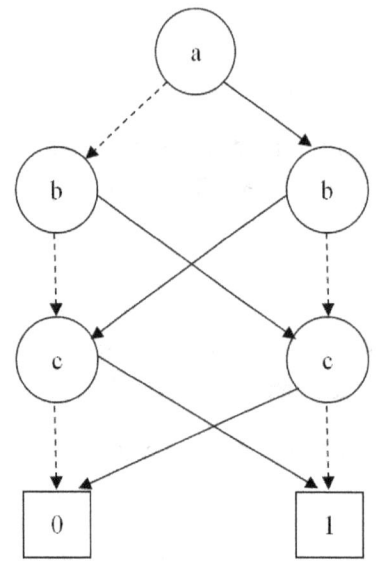

Figure 6.3 ROBDD for f = a⊕b⊕c

The ROBDD of Fig. 6.3 is of canonical form and small in size.

An ROBDD for f = a'b'c can be derived as follows from the binary tree.

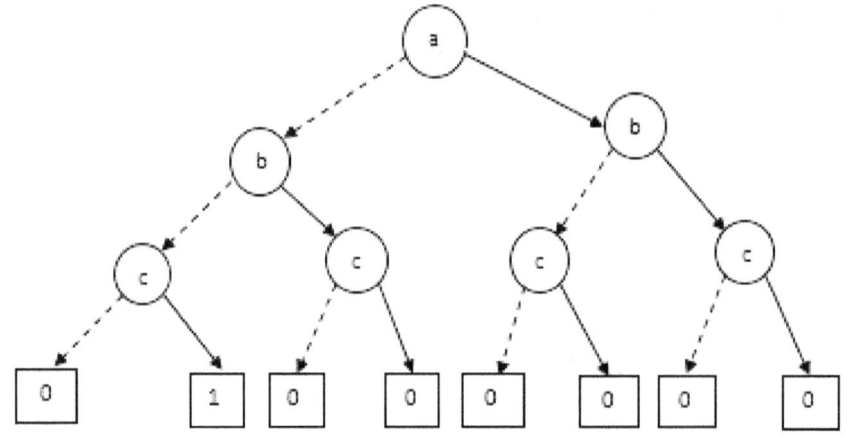

f = a'b'c above is equal to

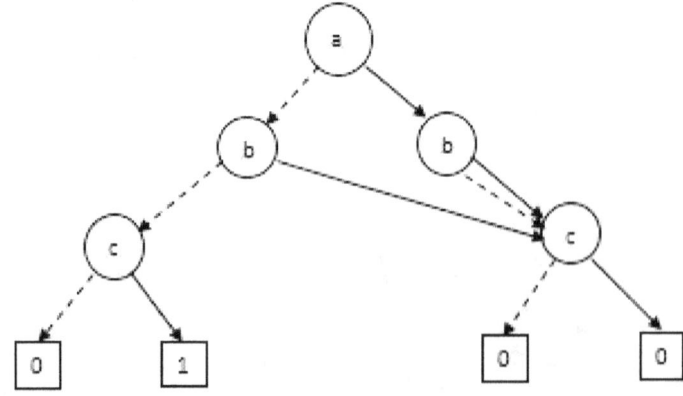

Figure 6.4

There are three nodes 'c' with identical terminal nodes, so, one of them can be retained as shown in Fig. 6.4 right side. The node 'b' on right side has two edges pointing to node 'c'. The node 'b' can therefore be eliminated as per rules enunciated earlier. This reduces to diagram on the left side of Fig. 6.5.

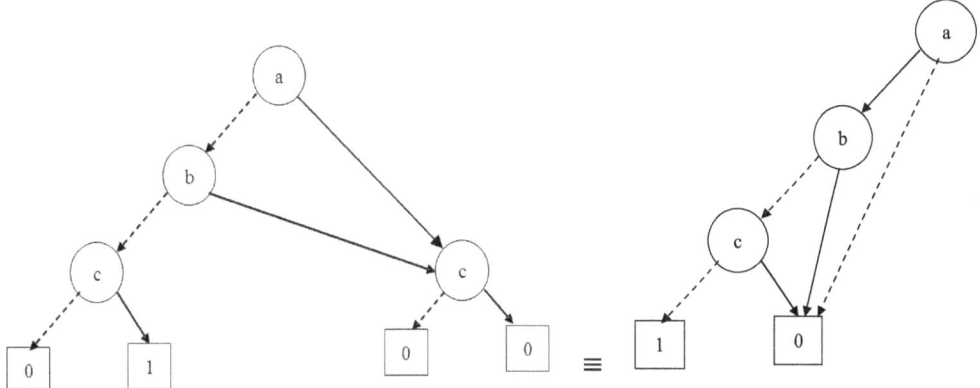

Figure 6.5

The node 'c' on the right side has two edges leading to '0's, this can be eliminated by directing the edges directly to '0' terminal node and the diagram on the left side is reduced to that on the right side of Fig. 6.5.

The BDD for f = ab'c' is derived similarly and is shown in Fig. 6.6

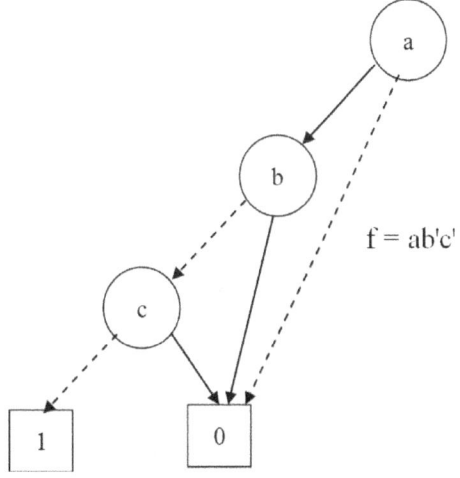

Figure 6.6

As per Bryant's procedures given in Fig. 6.7, two ROBDDs for a'b'c and ab'c' are added. BDDs are 'ANDed' or 'ORed' with terminal nodes.

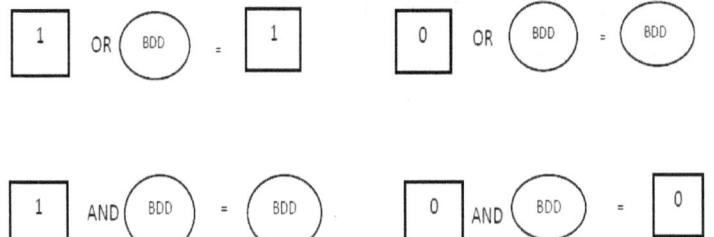

Figure 6.7

The ROBDD for $f = a'b'c + ab'c'$ is given in Fig. 6.8 and 6.9 after adding based on these procedures simplifying and eliminating duplicate nodes.

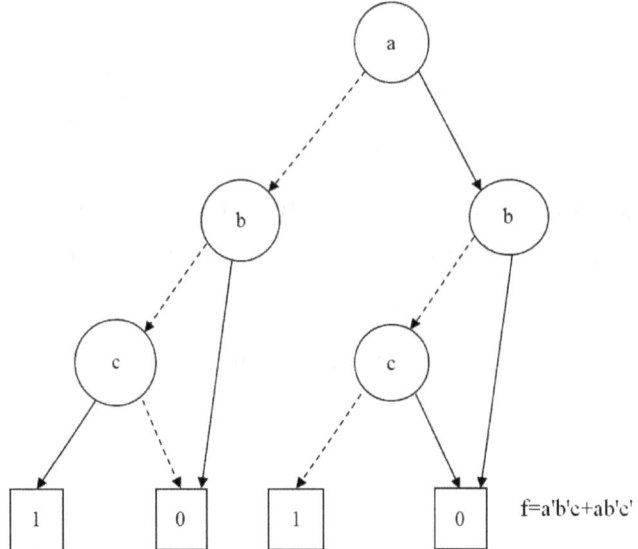

$f = a'b'c + ab'c'$

Fig 6.8

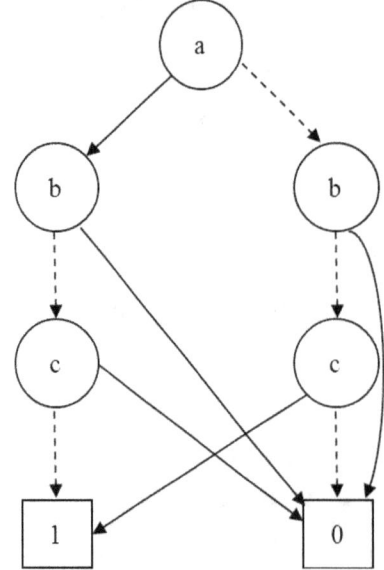

$f = a'b'c + ab'c'$ (after eliminating duplicate terminal nodes)

Figure 6.9

Similarly for f = a′bc′ + abc the ROBDD is given in Fig. 6.10.

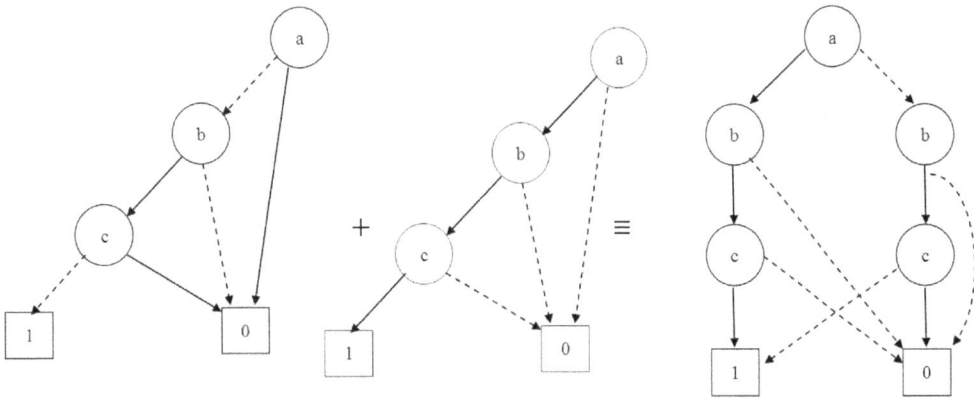

$$+ \qquad \equiv$$

f = a′bc′ + abc

Figure 6.10

Combining the results in Fig. 6.9 and Fig. 6.10 for the function f = a′b′c + ab′c′ + a′bc′ + abc the result is that given in Fig. 6.11.

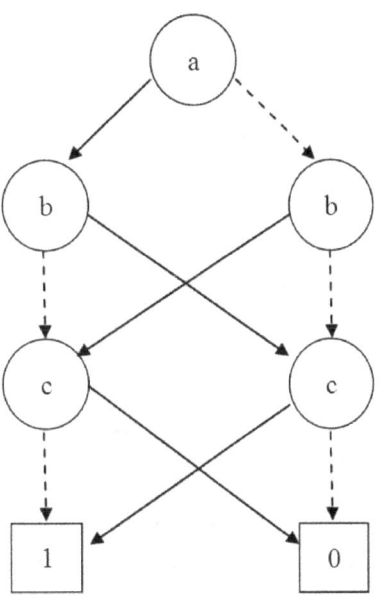

f = a′b′c + ab′c′ + a′bc′ + abc

Figure 6.11

6.2.1 It is found that the ROBDDs for f = a⊕b⊕c and f = a′b′c + ab′c′ + a′bc′ + abc shown in Fig. 6.3 and Fig. 6.11 respectively are identical. This is the basis of verification of an implemented circuit by comparing it to a specified circuit and if the ROBDDs are identical, then C≡R where C is the given circuit and R is the reference circuit.

6.2.2 For satisfiability for each output C_i and R_i, $C_i \oplus R_i = 0$. This is exemplified by the general formulation of the satisfiability problem as shown in Fig. 6.12.

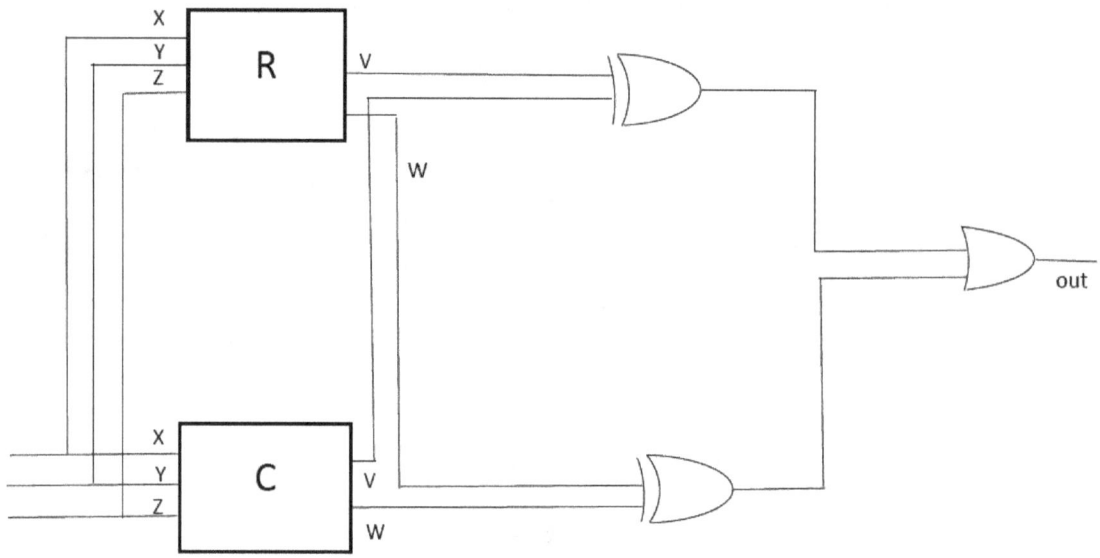

General Formulation of the Satisfiability Problem
C is identical to R only when out = = 0
(no inputs x, y, z that makes out = = 1)

Figure 6.12

6.3 Verification by Simulation

As already mentioned in paragraph 6.1.3, simulation is carried out for verification of medium size circuits and parts of larger circuits. For larger circuits including sequential forms, formal methods involving temporal logic models will have to be employed. Simulation is carried out by means of an accurate description of the hardware through Hardware Description Languages (HDL) either Verilog HDL or VHSIC HDL or VHDL. The hardware circuit is expressed in the form of a computer programming language which is compiled and executed on a computer. The output values as well as waveforms generated by the computer are verified to see that no design errors have cropped in. Verilog HDL is generally preferred as it is similar to the 'C' language and there are also translators which can translate 'C' or 'C++' to Verilog HDL or VHDL.

6.3.1 Example of an up counter

Verilog HDL program for an up counter including the test bench is shown in Fig. 6.13. This is simple behavioural modelling in Verilog [3], the period of the counter pulse is 8 ps and every 4 ps the sign changes. The counter counts till N=12 as programmed and resets to zero. The output count value 'a' starting from zero is incremented at every negative edge of the clock till it reaches the count N. The simulation is carried out through ModelSim–Intel FPGA Starter Edition 10.5b. The screenshot of the simulated waveform is shown in Fig. 6.14. The simulation is run upto 200 ps as seen.

```
module counterup (a, clk, N);
input clk;
input[3:0]N;
output[3:0]a;
reg[3:0]a;
initiala=4'b0000;
always@[negedgeclk)a=(a==N)?4'b0000:a+1'b1;
endmodule

module tst_counterup;//TEST_BENCH
reg clk;
reg[3:0]N;
wire[3:0]a;
counterupc1(a, clk, N);
initial
begin
        clk=0
        N=4'b1100;
end
always#4 clk=~clk;
initial $monitor($time,"a=%b, clk=%b, N=%b", a, clk, N);
endmodule
```

Figure 6.13 Verilog Module of an Up Counter

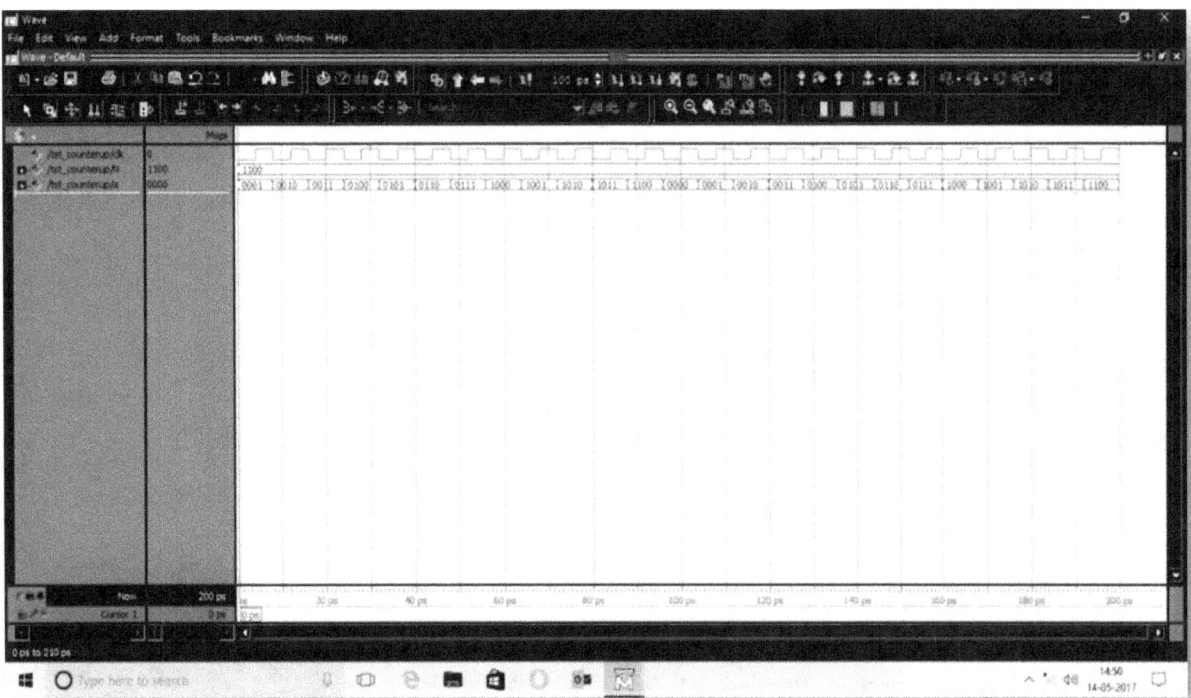

Figure 6.14 Screenshot of waveform of Up counter

```
module shifriter(a, clk, r_l);
input clk, r_l;
output [7:0]a;
reg[7:0]a;
initial a=8'h01;
begin
always@(negedge clk)a=(r_l)?(a>>1'b1):(a<<1'b1);
end
endmodule

module tst_shifriter;//test_bench
reg clk, r_l;
wire [7:0]a;
shifrlter shrr(a, clk, r_l);
initial
begin
 clk=1'b1;
   r_l=0;
end
always #4 clk=~clk;
initial #16 r_l=~r_l;
initial
$monitor($time,"clk=%b, r_l=%b, a=%b", clk, r_l, a);
initial #32 $stop;
endmodule
```

Figure 6.15 Verilog Module for Shift Register

```
# Reading C:/Users/dell/Documents/modelsim_ase/tcl/vsim/pref.tcl
Vlog-reportprogress 300-work work C:/Users/dell/Documents/shiftregister.v
# Model Technology ModelSim-Intel FPGA Edition vlog 10.5b Compiler 2016.10 Oct 5 2016
#Start time: 19:47:17 on May 16, 2017
#vlog-reportprogress 300-work C:/Users/dell/Documents/shiftregister.v
#--Compiling module shifriter
#--Compiling module tst_shifriter
#
#Top level modules:
#        tst_shifriter
# End time: 19:47:17 on May 16, 2017, Elapsed time : 0:00:00
#Error: 0, Warnings : 0
vsim-gui work.tst_shifriter
#vsim-gui work.tst_shifriter
#Start time: 19:47:48 on May 16, 2017
#Loading work.tst_shifriter
#Loading work.shifriter
Run
#        0clk=1, r_l=0, a=00000001
#        4clk=0, r_l=0, a=00000010
#        8clk=1, r_l=0, a=00000010
#        12clk=0, r_l=0, a=00000100
#        16clk=1, r_1=1, a=00000100
#        20clk=0, r_l=1, a=00000010
#        24clk=1, r_1=1, a=00000010
#        28clk=0, r_l=1, a=00000001
#**Note: $stop : C/Users/dell/Documents/shiftregister.v(24)
#        Time: 32 ps Iteration: 0 Instance:/tst_shifriter
#Break in Module tst_shifriter at C:/Users/dell/Documents/shiftregister.v line 24
# End time: 20:06:57 on May 16, 2017, Elapsed time: 0:19:09
# Errors: 0, Warnings: 0
```

Figure 6.16 Simulation of Shift Register – output

6.3.2 Simulation of a shift register

The component circuits of the microprocessor and the peripherals in the electronic interlocking system should be thoroughly tested in a physical laboratory and equivalent circuits in respect of inputs and outputs should be derived. These circuits should be simulated by means of HDL tools either verilog or VHDL. A shift register is part of the system and a verilog module including test bench is shown in Fig. 6.15. The register shifts by one bit to the right if r_l = 1 and to the left by one bit otherwise (i.e. if r_l = 0). Every 4 ps, the clock(clk) reverses and the r_l control signal is steady for 0-16 ps as can be seen in the output depicted in Fig. 6.16.

6.3.3 Simulation of an ALU

An important part of a microprocessor is the Arithmetic/Logic Unit or simply the ALU. A simple circuit of the ALU with two arithmetic functions and two logic function is simulated with the Verilog module depicting the behavioural functions shown in Fig. 6.17. The arithmetic and logic functions are selected by the control signal 'f' as shown. The program stops at 10 ps after the four functions are simulated every 2 ps. The simulated output is shown in Fig. 6.18. The output 'c' is seen after the activation of each of the four arithmetic/logic functions.

6.3.4 Simulation of a faulty ALU

After the physical testing and drawing of the equivalent circuits from the behavioural aspects, artificial faults can be introduced and the modules simulated incorporating the faults. The output results will indicate whether the concerned circuits produce the desired outputs. In the ALU under consideration, an artificial fault of the control signal 'f' being stuck at '10' has been introduced and the Verilog program of the faulty ALU module(fyalu) is shown in Fig. 6.19. The program is run upto 12 ps and the result of simulation is shown in Fig. 6.20. It is seen that the output 'c' gives appropriate values for the stuck fault. For example, when 'f' is stuck at '10' the circuit operates permanently as an AND circuit, say, when a = 1100 and b = 1101, c = 1100 appropriately.

6.3.5 Similarly the components of the controlling microprocessor and the peripherals can be simulated and tested with induced faults. A VHDL or Verilog HDL model of the CPU can be generated and simulated faults can be checked. Such a method of verification was adopted by Battelle Memorial Institute Columbus Ohio USA for the Microlok System of US&S.

6.4 Non-functional Hardware Testing

In addition to the various forms of functional testing including black box testing which comprises testing of hardware and software simultaneously, it is necessary that the hardware is tested for analog and digital parameters, stress and environmental factors, electro-magnetic compatibility etc.

```
module alubeh(c, s, a, b, f);
output[3:0]c;
output s;
input[3:0]a, b;
input[1:0]f;
reg s;
reg[3:0]c;
begin
always@(a or b or f)
        case(f)
        2'b00:c=a+b;
        2'b01:c=a-b;
        2'b10:c=a & b;
        2'b11:c=a| b;
        endcase
end
endmodule


module tst_alubeh;//test_bench
reg[3:0]a, b;
reg[1:0]f;
wire[3:0]c;
wire s;
alubeh aa(c, s, a, f);
initial
begin
f=2'b00;a=4'b0000;b=4'b0000;
end
always
begin
        #2 f=2'b00;a=4'b0011;b=4'b0000;
        #2 f=2'b01;a=4'b0001;b=4'b0011;
        #2 f=2'b10;a=4'b1100;b=4'b1101;
        #2 f=2'b11;a=4'b1100;b=4'b1101;
end
initial $monitor($time, "f=%b, a=%b, b=%b, c=%b,", f, a, b, c);
initial #10 $stop;
endmodule
```

Figure 6.17 Verilog Module including test bench for ALU

#Reading C:/Users/dell/Documents/modelsim_ase/tcl/vsim/pref.tcl

Vlog-reportprogress 300-work work{C:/Users/dell/Documents/MODULE ALU.v}

#Model Technology ModelSim-Intel FPGA Edition vlog 10.5b Compiler 2016.10 Oct 5 2016

#Start time: 20:16:06 on May 20, 2017

#vlog-reportprogress 300-work work C:/Users/dell/Documents/MODULE ALU.v

#--Compiling module alubeh

#--Compiling module tst_alubeh

#

#Top level modules:

tst_alubeh

#End time:20:16:07 on May 20, 2017, Elapsed time:0:00:01

#Errors: 0, Warnings: 0

vsim-gui work.tst_alubeh

#vsim-gui work.tst_alubeh

#Start time:20:16:39 on May 20, 2017

#Loading work.tst_alubeh

#Loading work.alubeh

run

0f=00, a=0000, b=0000, c=0000,

2f=00, a=0011, b=0000, c=0011,

4f=01, a=0001, b=0011, c=1110,

6f=10, a=1100, b=1101, c=1100,

8f=11, a=1100, b=1101, c=1101,

#**Note:$stop : C/Users/dell/Documents/MODULE ALU.v(37)

Time:10 ps Iteration:0 Instance:/tst_alubeh

#Break in Module tst_alubeh at C:/Users/dell/Documents/MODULE ALU.v line 37

Figure 6.18 Simulation of ALU-output

```
module fyalu(c, s, a, b, f);
output[3:0]c;
output s;
input [3:0]a, b;
input[1:0]f;
reg s;
reg[3:0]c;
begin
always@(a or b or f)
        case(f)
        2'b00:c=a+b;
        2'b01:c=a-b;
        2'b10:c=a&b;
        2'b11:c=a|b;
        endcase
end
endmodule

module tst_fyalu;//test_bench
reg[3:0]a, b;
reg[1:0]f;
wire[3:0]c;
wire s;
fyalu aa(c, s, a, b, f);
initial
begin
f=2'b00;a=4'b0000;b=4'b0000;
end
always
begin
        #2 f=2'b10;a=4'b0011;b=4'b0000;
        #2 f=2'b10;a=4'b0001;b=4'b0011;
        #2 f=2'b10;a=4'b1100;b=4'b1101;
        #2 f=2'b10;a=4'b1100;b=4'b1101;
end
initial $monitor($time, "f=%b, a=%b, b=%b, c=%b", f, a, b, c);
initial #12 $stop;
endmodule
```

Figure 6.19 Verilog Module including test bench for faulty ALU

```
# Reading C:/Users/dell/Documents/modelsim_ase/tcl/vsim/pref.tcl
vlog-reportprogress 300-work work C:/Users/dell/Documents/fyalu.v
# Model Technology ModelSim – Intel FPGA Edition vlog 10.5b Compiler 2016.10 Oct 5 2016
# Start time : 19:50:46 on May 18, 2017
# vlog-reportprogress 300-work work C:/Users/dell/Documents/fyalu.v
# -- Compiling module fyalu
# -- Compiling module tst_fyalu
#
#Top level modules:
#          tst_fyalu
#End time: 19:50:47 on May 18, 2017, Elapsed time: 0:00:01
#Errors: 0, Warnings: 0
vsim-gui work.tst_fyalu
#vsim-gui work.tst_fyalu
#Start time: 19:51:12 on May 18, 2017
#Loading work.tst_fyalu
#Loading work.fyalu
Run
#               0f=00, a=0000, b=0000, c=0000
#               2f=10, a=0011, b=0000, c=0000
#               4f=10, a=0001, b=0011, c=0001
#               6f=10, a=1100, b=1101, c=1100
#               10f=10, a=0011, b=0000, c=0000
#**Note: $stop : C:/Users/dell/Documents/fyalu.v(37)
# Time: 12 ps Iteration: 0 Instance: /tst_fyalu
#Break in Module tst_fyalu at C:/Users/dell/Documents/fyalu.v line 37
```

Figure 6.20 Simulation of faulty ALU – output

6.4.1 Under the 'Standardisation Testing and Quality Certification' (STQC) Directorate of the Government of India, there are Regional Test Laboratories and Electronic Test and Development Centres located in the four regions of India. These organizations provide testing facilities for the industries and other clients. The testing includes (1) component testing (2) environment testing (3) EMC (Electro Magnetic Compatibility) testing (4) system testing (5) reliability testing and analysis (6) safety testing etc. The laboratories are equipped with the state of the art testing facilities and offer assurance of compliance to technical specifications both national and international. The quality compliance is done with respect to ISO/IEC 17025.

6.4.2 The laboratories of the STQC Directorate are approved by the Ministry of Railways/RDSO for testing needs and compliance testing to RDSO specifications RDSO/SPN/192/2005 and RDSO/SPN/144/2014 concerning EI can be easily carried out.

6.4.3 There are also specialized organizations under STQC for example the CFR (Centre For Reliability) in Chennai which provides state of the art services in Availability, Reliability, Maintainability and Safety. If has facilities for HALT (Highly Accelerated Life Testing) which is essential for longevity of electronic components. Probability Ratio Sequential Test (PRST) for reliability demonstration can also be carried out in this institute.

6.5 Application of HOL to verification of hardware

Higher Order Logic or HOL is being applied for verification of complex circuits, as the older hardware verification techniques can be applied only to small and medium sized circuits with emphasis on controllers and combinational circuits [5]. Also the most abstract representative view is gate level. HOL enables the application of abstraction techniques and the use of abstract data types including usage of simple natural numbers. Also hierarchical verification approaches can be easily used. The logic is very expressive but the proofs cannot be automated at present, will have to depend on proof strategies and modelling techniques.

6.5.1 Syntax of Higher Order Logic (HOL)

As readers may not have a good knowledge of theoretical computer science, the terms in the syntax of HOL are being listed in a simple form. The terms in the syntax comprise (1) constants (2) variables (3) function applications (4) (lambda) abstractions (5) truth values (6) negations (7) and – connections (8) or – connections (9) implications (10) equivalences (11) equalities (12) universal quantifiers (13) existential quantifiers (14) unique existences (15) Hilbert's choice operators (16) conditionals. The symbols in all the terms are omitted for simplicity.

6.5.2 Abstraction Mechanisms

Abstraction can be performed in four different ways:

Structural abstraction: The 'black box' behavior is only considered i.e., the values at the primary inputs and outputs, not the values in the internal circuits.

Behavioural abstraction: A certain part of the overall behavior is considered for example, a certain safety property.

Data abstraction: The behavior is considered in terms of abstract data types, for example, using natural numbers instead of boolean values.

Time abstraction: The specified behavior is considered in a coarser time scale in comparison to the implementation time scale, for example, in terms of instruction cycles instead of nanoseconds.

6.5.3 Example of a one bit full adder

The one bit full adder as shown in Fig. 6.21 has the specification given in paragraph 6.2.

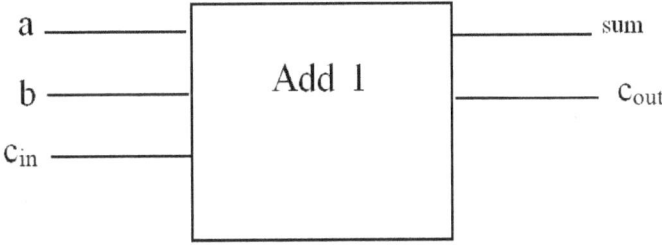

Figure 6.21 One bit full adder

$Add1\_spec$(a: $\mathbf{N}$, b:$\mathbf{N}$, $_{cin}$:$\mathbf{N}$, sum:$\mathbf{N}$, c_{out}:$\mathbf{N}$):=

$2.c_{out} + sum = a + b + c_{in}$ -- (6.2)

(**N** indicates natural numbers in HOL)

The implementation of the one bit full adder is given in Fig. 6.22.

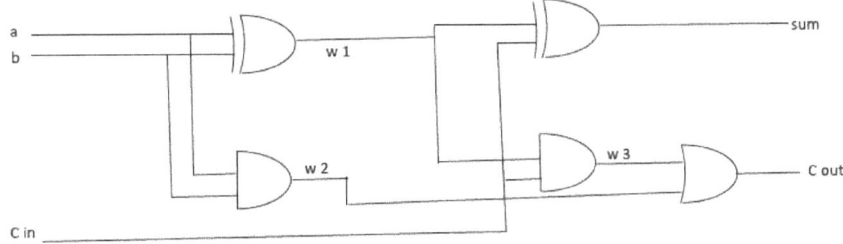

Figure 6.22 Implementation of one bit full adder

The implementation as given in Fig. 6.22 can be depicted in the following equation 6.3.

$Add1\_Imp$ (a:$\mathbf{B}$, b:$\mathbf{B}$, C_{in}:$\mathbf{B}$, Sum:$\mathbf{B}$, C_{out}:$\mathbf{B}$):=

$\exists w_1 w_2 w_3$. $Exor$(a, b, $_{w1}$)$\wedge$
And(a, b, $_{w2}$)$\wedge$
$Exor$(w_1, $_{cin}$, sum)$\wedge$
And(w_1, $_{cin}$, $_{w3}$)$\wedge$
Or(w_2, $_{w3}$, $_{cout}$) --- (6.3)

(**B** – Boolean, $\exists$ – there exists, $\wedge$ – and)

To formulate a useful proof goal, a data abstraction function BV:$\mathbf{B}\rightarrow\mathbf{N}$, mapping Boolean values to natural numbers is required. In this case, the definition in (6.4) below suffices.

$$BV(x) := \begin{cases} 1, & if\ x = T \\ 0, & if\ x = F \end{cases} \quad \text{---} \quad (6.4)$$

Based on the above definition, the equivalence needed for successful verification can be expressed as

$Add1\_Imp(a, b, c_{cin}, sum, c_{out}) =$

$Add1\_Spec(BV(a), BV(b), BV(c_{in}), BV(sum), BV(c_{out}))$ ------------------------------------- (6.5)

The above equation can be proved theoretically by expanding the predicates and substituting with natural numbers. This type of proof has been handled in different ways depending on the type of HOL chosen. There are different HOLs in the field, some of which are HOL98, Isabelle HOL, Lambda HOL, PVS HOL, Veritas HOL, HOL Light etc. Frequently the proof is built over generating an algorithm and automating it to get the result.

6.5.4 Application of HOL

HOL has been applied to verify complex circuits in microprocessors by many hardware manufacturers. HOL Light has been used by J. Harrison of Intel Corporation [2] to verify the Floating Point Unit (FPU) of Intel microprocessor. D. Russinoff has verified the correctness of FPU of AMD Athlon processor by employing HOL (ACL2).

References

[1] Gaetano Borriello "Verification with BDDs" – lecture notes #23 of Prof. G. Borriello University of Washington Seattle – 2008.

[2] J. Harrison "HOL Light: an overview" – Intel Corporation – 2006.

[3] T.R. Padmanabhan et al. "Design through Verilog HDL" – Wiley India 2015.

[4] Paul Jackson "Introduction to Binary Decision Diagrams (BDDs)" University of Edinburgh – 18 Nov 2013.

[5] Thomas Kropf "Introduction to Formal Hardware Verification" Springer – Verlag 1999.

## CHAPTER 7

VERIFICATION AND VALIDATION OF SOFTWARE

7.1 Objectives and tasks

The objective of software verification and validation throughout the life cycle, is to ensure quality and compliance with the user requirements [7]. The *verification* is done to check whether the already laid down requirements are met and *validation* is done to ensure that the user requirements are satisfied. During this process, the software is changed or altered to meet the requirements and all the changes have to be properly documented.

7.1.1 The software is subjected to review, analysis and testing to verify the compliance to requirements which include functional as well as quality aspects. The V & V (verification & validation) effort is applied usually in parallel with the software development and the result of the V & V as communicated to the developers will assist in maintaining the functionality and quality at each stage till the software components are integrated and reviewed. The V & V tasks are performed either by a separate group in the developer organization or an independent organization.

7.1.2 V & V tasks during the lifecycle

The general tasks during V & V are *(1) traceability analysis (2) evaluation (3) interface analysis* and *(4) testing*[7]. These are described in the following paragraphs.

7.1.2.1 Traceability analysis

Traceability refers to the linkage between the original requirements and the subsequent phases in the life cycle such as design, testing, implementation etc. The forward and backward linkage between the elements of the various phases aids in verification of the properties set in the requirements. It is verified whether the properties set at the concept stage are transferred to the design specifications, to the coding and testing stages and finally to the system to be delivered. Also traceability analysis helps in evaluating the software development effort for good software engineering practices.

7.1.2.2 Evaluation

The developed software is evaluated to see whether it is fit for use, confirms to specifications, standards and quality attributes. The evaluation is done through (1) static analysis-reviews, inspections etc (2) dynamic analysis-simulation, prototyping etc. and (3) formal analysis-mathematical proof. Dynamic analysis includes execution of the software code in some manner, hence gives better results and helps in determining the corrections and modifications to be done in lesser time than the other methods. Formal analysis is very much useful in verifying safety critical software and increases the confidence placed in the requirements.

7.1.2.3 Interface analysis

Interface analysis is concerned with data flows between any two parts of a system and it is to ensure that the data is complete, accurate and consistent when transmitted over the interfaces. Focus should be on three interface areas namely (1) user interface-the format in which the information is presented to the user (2) hardware interface-identify all the hardware devices for the interfaces and the applicable standards and (3) software interface-identify interfaces to other software products, such as, data management system, operating system etc and the applicable standards. The specification for data and message formats, word length, general and communication protocols etc. have to be looked into for successful functioning of the interfaces.

7.1.2.4 Testing

Testing is the process of physical verification whether the objectives have been met in the software under development and is an important part of the V & V process. The testing is performed during the life cycle at various stages and the testing can be subdivided into four stages namely *(1) component or unit testing*-each unit or module or subsystem is tested and the compliance to design is verified *(2) Integration testing*-hardware and software elements in each unit are combined and tested till full integration *(3) system testing*-the integrated hardware and software system is tested to verify the specified requirements *(4) Acceptance testing*-formal testing is done to verify whether the acceptance criteria are satisfied and to enable the customer to accept it. Organizational aspects and detailed test plans are laid down for each stage in the life cycle process.

7.1.2.5 Details of tests before acceptance for safety critical systems

Rigorous testing at the unit level and system level are required before a safety critical system is accepted by the user [5]. As per the European experience 20 to 50 errors per 1000 lines of code are found white testing during development, still 1.5 to 4 errors per 1000 lines remain even after system testing, only to be rectified while in early stages of use. The detailed tests for each of the four stages are described below.

(a) Unit tests

The unit in Indian parlance is the card or PCB (Printed Circuit Board) and a number of PCBs connected over the back plane constitute the rack. Again as per European experience, 65% of bugs can be caught in unit testing and half of these can be found out during 'white-box' tests. Unit testing is considered vital in removing the bugs as less software and hardware are involved. Three tests are performed at unit level i.e. *(1) white box test (2) black box test and (3) performance test.* In the *white box test,* each branch in the logic is tested and the preferred method is the structural method based on cyclomatic complexity. This is a method commonly described in 'software engineering' text books, hence not detailed here. Either debugging tools or diagnostic codes are used to test the units. In *black box tests* also called function tests, the outputs are examined for each input to the unit taking the unit as a black box. As it is impractical to test all the possible inputs, the inputs are divided into equivalence classes, where inputs in each class produce the same error. In the *performance test,* the performance of the module is observed, such as, execution time or CPU time, amount of usage of memory etc. by means of performance analysis tools, diagnostic code and system monitoring tools.

(b) Integration tests

Two or more units are combined to make a subsystem depending on the configuration of the system and these units must have a common control flow. The three tests conducted on the individual units namely, white box test, black box test and performance test are repeated for the subsystem. The subsystems are selected such that the whole system is covered. Interface tests covering the interfaces between the subsystems are conducted choosing the functional aspects.

(c) System tests

System tests which are essential for safety critical systems are listed here:

(1) *Function tests:* Each functional requirement is tested similar to a black box test feeding inputs and examining outputs.

(2) *Performance tests:* system performance parameters as listed have to be checked and worst case performance targets are verified.

(3) *Interface tests:* Conformance to external interface requirements including human interfaces has to be verified. Simulators and other test tools may be required if the tests are to be conducted outside the operational environment.

(4) *Maintainability tests:* Mean Time To Repair (MTTR) is estimated by taking the average of the times to repair recorded while testing. It is to be checked whether this figure is reasonable, if not, the size and complexity of modules have to be altered. including the code employed.

(5) *Safety tests:* System behaviour will have to be observed while causing deliberate faults. For exhaustive testing, simulators may have to be built.

(6) *Stress tests:* The performance of the system is checked when it works beyond the limits of its specified requirements, for example, the maximum number of activities that can be supported simultaneously and maximum quantity of data that can be processed in a given time. These may be very relevant in case of communication protocol based control systems.

(d) Acceptance tests

The system is tested to verify the user requirements with the software already successfully tested at system level. Acceptance tests are to be conducted by the user or his representative. Most of the tests desired by the user are a replica of the system level tests and the system is accepted by the user, if the tests at system level conducted by the supplier are certified to be successful.

7.2 Choice of computer languages

Many language analysts such as the authors of [4] have not recommended 'C' language at all for safety-critical systems due to the factors (1) wild jumps (2) overwrites (3) problems in integer overflow (4) data typing (5) exception handling and (6) problems in storage overflow etc. However 'C' and 'C++' are used in safety critical systems with safety constraints, because of better control over memory management, simple well debugged core runtime libraries and mature tool support. 'C' language is used extensively in the automobile industry and in many European systems whereas 'Ada' is preferred in North America for safety critical systems. As per table A.15 of EN 50128, 'C' and 'C++' languages are only 'R-Recommended' whereas ADA and PASCAL are 'HR-Highly Recommended'. However a subset of 'C' has been used extensively in Europe and other countries. As per D.35 of EN50128, a language subset is chosen to reduce the probability of introducing programming faults and increase the probability of detecting any remaining faults.

7.2.1 Features of languages to be avoided

As per D.54 of EN50128, the following features should be avoided in programming languages. (1) unconditional jumps excluding subroutine calls (2) recursion (3) pointers, heaps or any type of dynamic variables or objects (4) interrupt handling at source code level (5) multiple entries or exits of loops, blocks or subprograms (6) implicit variable initialization or declaration (7) variant records and equivalence and (8) procedural parameters.

7.3 Good programming practice

Validation of software will become easier if good programming practice is followed and the software is well structured, understandable, readable, printable and reusable. Some of the steps for good programming are listed below [14].

7.3.1 Modularization: If different programs perform identical tasks, the identical operations should be collected in common modules. Such modules (stored in libraries) are easier to maintain and safer to use than inserted copies of identical source code. Large programs can be broken into separate, smaller programs or modules which can be separately specified, written and tested. The smaller modules make it easier to understand the problem and this reduces the scope for error and eases the task of checking. The optimum size of the module depends on the function of the module and as a rule of thumb, the modules should not normally exceed 100 lines of code in HLL (High Level Language) [13].

7.3.2 Structured Programming: This is a method of using certain clear and well defined approaches to program design instead of complex programs which are difficult to understand or inspect and are prone to errors. A major source of error in programs is the use of the GOTO statement for constructs such as loops and branches (decisions). The structured programming approach discourages the use of GOTOs, requiring the use of control structures which have a single entry and a single exit [13]

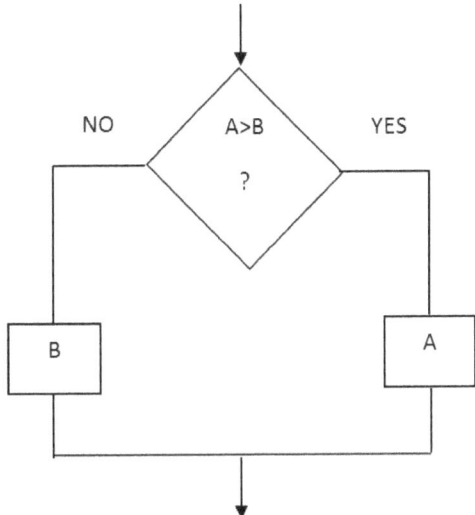

Unstructured: if A>B go to A (line number), else go to B (line number)

Structured: if A>B then A (subroutine), else B (subroutine)

Fig. 7.1 Structured vs. unstructured programming

7.3.2.1 A simple branch instruction shown in Fig. 7.1 can be programmed (in 'Basic') in either an unstructured or a structured way. The unstructured approach can lead to errors if the wrong line number is given (or if line numbers are changed as a result of program changes) and it is difficult to trace the subroutines (A, B) back to the decision point. On the other hand, the structured approach eliminates the possibility of line number errors and is much easier to understand and inspect. The drawback of structured programming is in terms of speed and memory requirements.

Functional division: Functionality should be broken down into small manageable and testable units. Often used operations and calculations should be isolated so that identical performances are executed by the same code.

Revision notes: Programming revisions and changes to released executable code should always be documented in the source code.

Naming conventions: Function and parameter should express their meaning and use.

Readable source code: Source code should be readable. Word-wrap in the text makes it difficult to read.

Fail - safe: The program should issue an error message whenever an error is detected and respond accordingly. Debugging options that can be used to catch run-time error conditions should never be used in released executable code.

7.4 Safe subset of 'C'

'C' language is extensively used in embedded systems as it gives good support for the high speed low-level input/output operations and can generate smaller and less RAM - intensive code than many other high-level languages. However for safety critical applications, constraints have to be set in programming the language to avoid possible errors in execution. The availability of many tools for development of 'C' has helped in extensive use of 'C' in the aerospace and automobile industries in Europe and other countries. Detailed rules including sub-divisions amounting to more than a hundred have been prescribed by the Motor Industry Software Reliability Association – MISRA-C for the use of 'C' in the automobile industry [11].

7.4.1 Some constraints which are considered more prominent are reproduced here[12]. Many more can be added, based on experience and necessity.

(1) The number and size of parameters passed to routines should be limited, as too many parameters affect readability and testability of the routine.

(2) Recursive functions should also be limited as stack overflows are common.

(3) Functions for boundary checking have to be created, as 'C' does not do this automatically.

(4) 'gets' and other related functions should not be used as these do not have adequate limit checks.

(5) If "*if...else if...else if...*" gets beyond two levels, use a "*switch...case*" instead, as this increases readability.

(6) When using "*switch...case*", *default* should be explicitly defined, '*break*' should normally be included.

(7) Local variables should be initialised, as they contain garbage before explicit initialisation.

(8) Pointers to functions should be avoided as these pointers cannot be initialised.

(9) ++ or – operators should not be used on parameters being passed to subroutines or macros, as these can create unexpected side effects.

(10) Interrupts should be avoided and if they are to be used, functions within interrupt service routines should not be called.

(11) Assumptions should not be made about the sizes of dependent types such as *int*. The size is often platform and compiler dependent.

(12) Library function names or reserved words should not be used as variable names, as this could lead to serious errors.

7.5 Suggestions for incorporation in IR specification RDSO/SPN/192/2005

As many bugs can be found in unit testing stage, by means of white-box testing, as per the information given by the European Space Agency [5], it is suggested that white-box testing can be included under 'Type Tests' in the specification RDSO/SPN/192/2005 (Appendix E). Though the number of card failures in the EI systems has been found to be small, it is likely that hidden failures can be found and rectified by the manufacturer at the development stage. The white-box tests can be conducted by the manufacturer and the results shown to the inspecting authority.

7.5.1 Software Engineering standards such as Software Capability Maturity Model-SW-CMM-level 3 should be specified in paragraph 8.2 of specification no. 192 for the supplier to follow.

7.5.2 Source code for the Executive software and Application software should be made available to RDSO and should be validated by the Independent Verification and Validation (IVV) authority. A list of firms authorised to perform IVV of software should be prepared by RDSO and given to the suppliers, so that there is no scope for entrusting the work to an unauthorised firm. Though paragraph 13.7 of RDSO/SPN/144/2014 (Appendix E) states that details of source code etc should be supplied to RDSO, this should also be mentioned in specification 192 and strictly enforced. The possession of source code will help in rectifying hidden faults if found during maintenance stage, after the supplier leaves the field.

7.5.3 It will be very much beneficial if a software engineering group is formed in RDSO itself for all the engineering departments, namely Civil, Mechanical, Electrical and Signal as many appliances employing varied source codes are being used in the field. This group can also supplement the validation of software done by the private firms.

7.5.4 The details of IVV including development tools used and simulator models developed by the firm performing IVV should be made available to RDSO. A software audit keeping in mind the quality assurance should also be performed by the firm.

7.6 Simulation of signalling and interlocking at a station

The signalling and interlocking at a station 'Y' has been simulated using 'C' language mostly and the associated graphics programs. Station 'Y' is largely based on the station yard at 'Kavaraipettai' on the Chennai – Gudur section of Southern Railway as existing in the 1980s. The station was extensively remodelled later. This yard is a simple double line station with a manually operated cross-over. The 'down' portion of the yard has been taken for simulation of the interlocking. TURBO C++ compiler has been utilized as it is the best for compiling graphics programs. The full program is listed in Appendix 'C'.

7.6.1 Route setting for the five routes possible on the 'down' yard have been simulated. These routes are clearly indicated in the table of control. The signalling plan and table of control for the 'down' yard are shown in the screenshot at Fig 7.2

7.6.2 The route '1A' is set by typing '*1A*' and 'ENTER'. The setting of the signals and the route '1A' are shown in the screenshot at Fig. 7.3. The 'gets (route)' function could not be avoided in the simulation but is not recommended for the safe subset of 'C' in actual operation as mentioned in paragraph 7.4.1. 'If…else' function has been avoided by choosing 'switch…case' function as the levels are more than two, complying with the constraint mentioned in paragraph (5) of 7.4.1

7.6.3 Simple logic has been used in simulation of the interlocking and two abnormalities i.e. simulating route 1B when route 1A is already set and simulating route 1A when route 1B is already set, are programmed through two functions abnorm 1( ) and abnorm 2( ). The simulation of abnorm 1( ) is shown in the screenshot at Fig 7.4. The screenshot shows the route 1A set and 'conditions not present for route 1B'. The activation is done by typing '1A *B' and 'ENTER'. Similarly for abnorm 2( ), it is '1B *A'.

7.6.4 The 'back locking' and 'approach locking' for route 1A have been simulated through '1ABL' and '1AAL' commands. The screenshots are not presented as it requires a 'video' to present all the sequences and it is beyond the scope here.

7.6.5 Signals for run through of a train have been simulated by the 'MLTH' command indicating mainline through reception. All signals are set at 'green', namely, the distant, home, starter and advanced starter, with the setting of routes 4F, 2E and 1A in that order.

7.7 Verification of 'C' programs with CBMC

CBMC or 'C bounded Model Checker' has evolved with the efforts of E.Clarke of Oxford University and D. Kroening of Carnegie Mellon University with grants from various bodies of the U.S. Government in the first decade of the 21st Century. The work was mostly carried out at CMU (Carnegie Mellon University) Department of Computer Science.

7.7.1 The CMBC software tool can be downloaded from *www.cs.emu.edu/~ model check/cbmc*. For Windows version of CBMC, the executive path is enabled through Microsoft Visual Studio, which is also to be downloaded. The CBMC and Microsoft Visual Studio (2015) have been downloaded by the author and verification of some 'C' programs is detailed below.

7.7.2 CMBC formally verifies ANSI-C programs and checks the properties such as array bounds, pointer safety, functional behaviour, division by zero and user provided assertions The program is translated into a set of bit vector equations and fed to a SAT (Satisfiability) solver. If there is no bug found, the verification is successful, if not, failure and wherever possible a counter example is produced. CBMC uses built-in modelling primitives (1) *xxx nondet_xxx( )* which non-deterministically returns a value of type *xxx* in an unknown environment (2) _ _CPROVER_ *assume* which restricts program traces to those satisfying the assumption (3) _ _ CPROVER_ *assert* enables verification of a statement or assertion and outputs 'failure' or 'success' as per the accompanying comment.

7.7.2.1 An example of verification of a simple program with undefined integer is illustrated. The program is given in Fig 7.5 based on a core program given in [6]. The program is verified by CBMC and the result is presented in the screenshot at Fig 7.6. The verification failed due to the undefined nature of function 'x'.

7.7.2.2 A program for verification of string array bounds taken from the simulation program for station 'Y'[Appdx C], with minor modifications is given in Fig. 7.7. This program is verified by CBMC and the screenshot of the result of verification is presented in Fig 7.8. As seen, the outcome is 'VERIFICATON SUCCESSFUL', as the arrays are well defined.

7.7.2.3 A program for the 'switch case' function is taken from the simulation program for station 'Y' at Appendix 'C'. The program with minor modifications to suit the verification is listed in Fig. 7.9. This is verified by CBMC and the result is shown in the screenshot at fig 7.10. The verification is 'SUCCESSFUL', also it has been compiled successfully by the 'TURBO C++' compiler and the results were already presented as screenshots in Figures 7.2, 7.3 and 7.4.

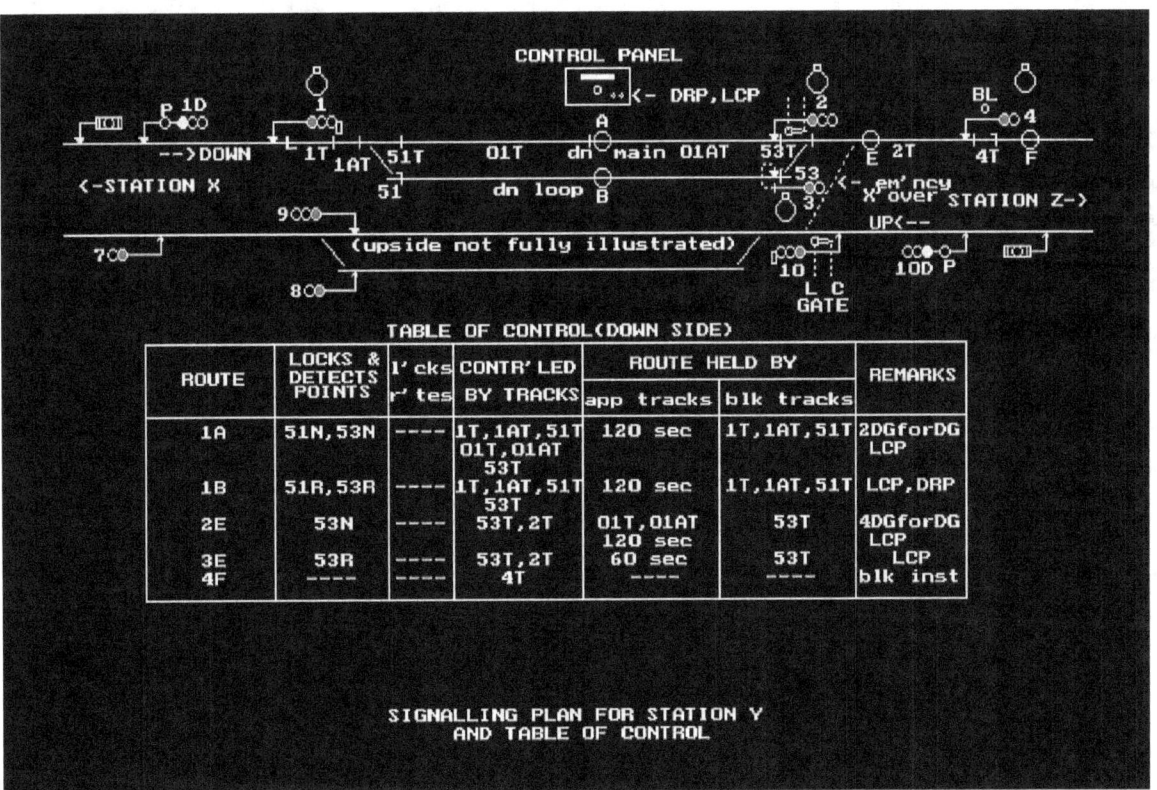

Fig. 7.2 Screenshot of signalling plan and table of control-station 'Y'

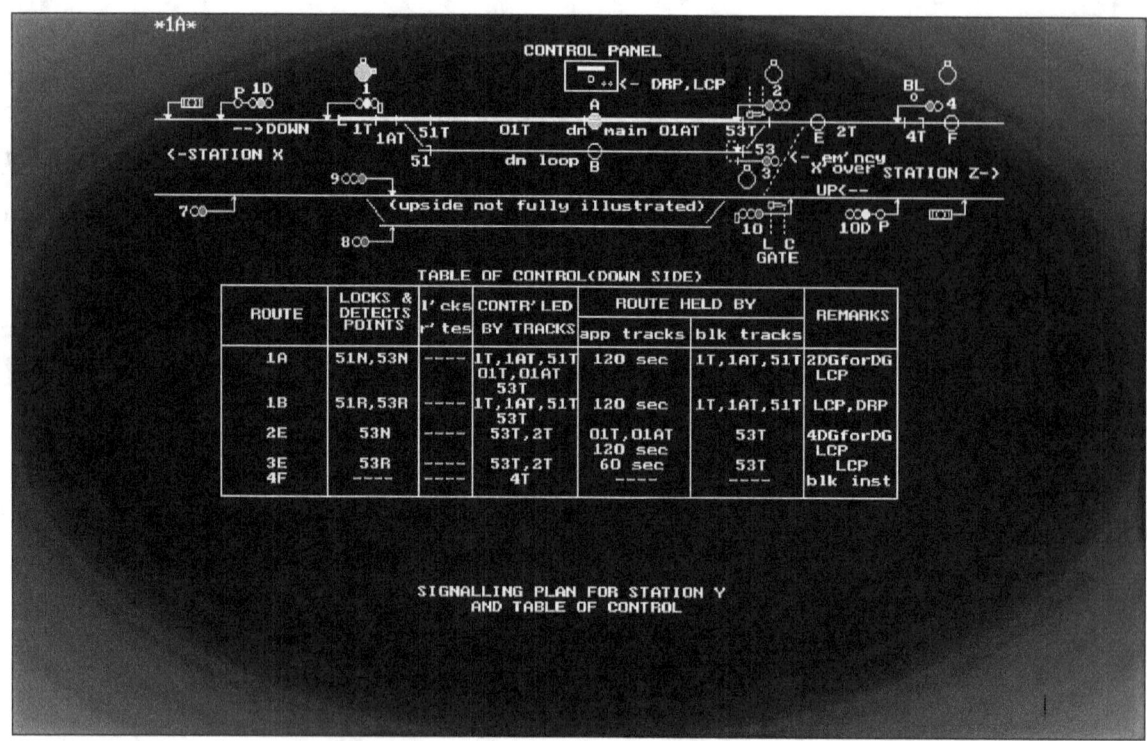

Fig 7.3 Screenshot of signalling with route 1A set

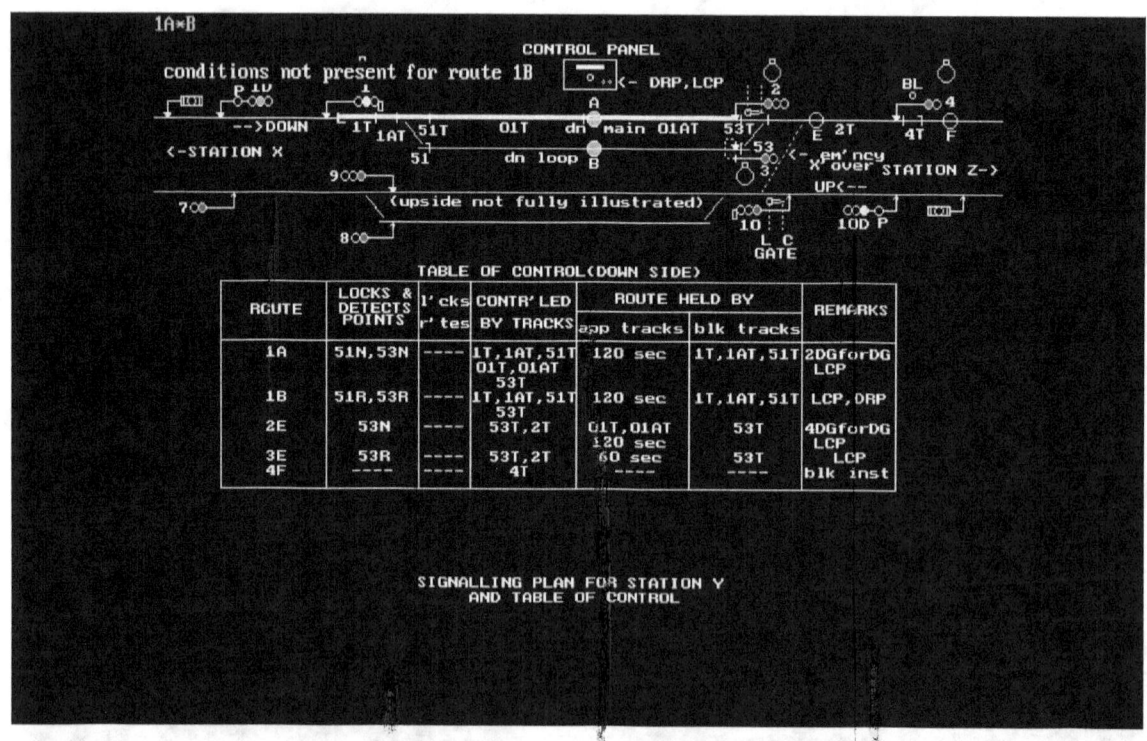

Fig 7.4 Screenshot of signalling not permitting route 1B with route 1A set

```
int x, y;
void main ()
{
x=nondet_int();
_CPROVER_assume(x>10);
Y=x+1;
_CPROVER_assert(y>x,"failure or success");
```

Fig 7.5 Simple program with undefined integer

Fig 7.6 Result of verification of the program in Fig 7.5 (Screenshot)

```
#include<stdio.h>
#include<conio.h>
#include<stdlib.h>
#include<dos.h>
#include<string.h>
main()
{
 char route 1[5]= "*1A*";
char route 2[5]= "*1B*";
}
```

Fig. 7.7 A simple program for checking array bounds

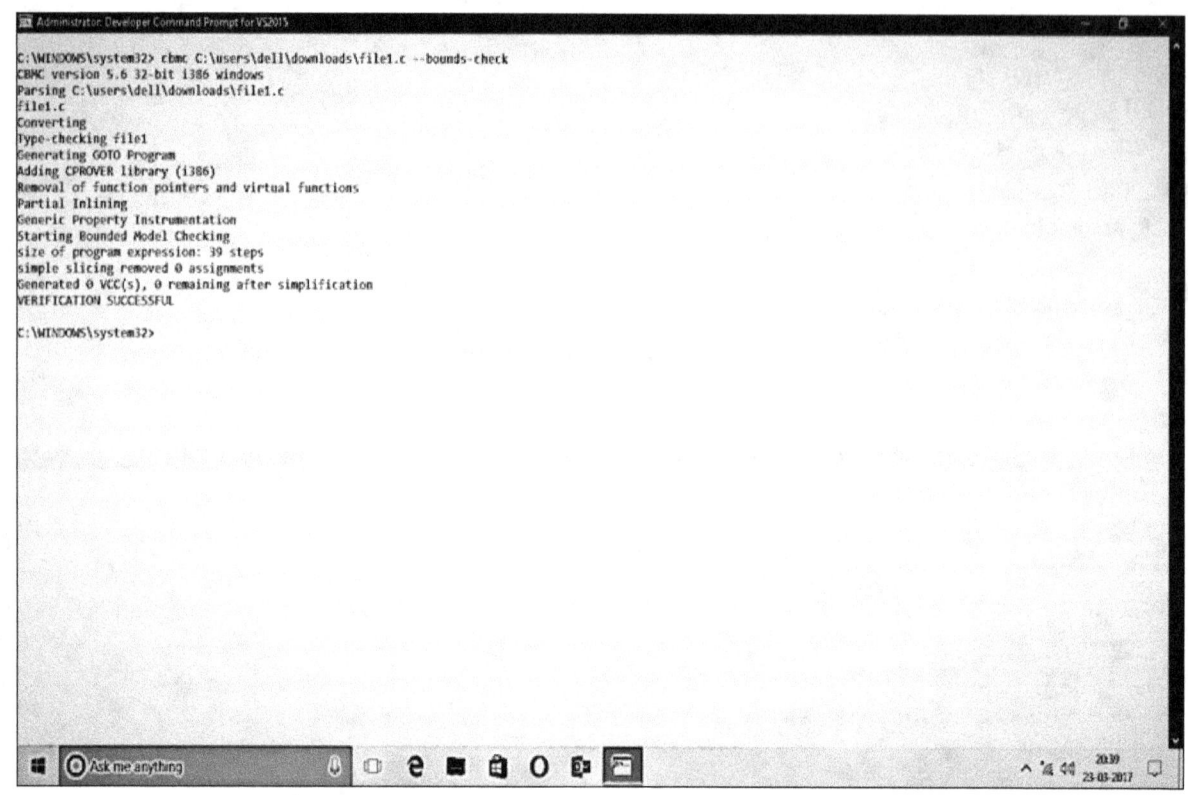

Fig 7.8 Result of verification of program in Fig 7.7 (screenshot)

```
#include<stdio.h>
#include<conio.h>
#include<stdlib.h>
#include<dos.h>
#include<string.h>
main( )
{
char route1[5]="*1A*";
char route2[5]="*1B*";
char route3[5]="*2E*";
char route4[5]="*3E*";
char route5[5]="*4F*";
char routechk1[5]="1A*B";
char routechk2[5]="1B*A";
char route1Abklock[5]="1ABL";
char route1Aapplock[5]="1AAL";
char route1Amlthrecep[5]="MLTH";
void setroute1A( );
void setroute1B( );
void setroute2E( );
void setroute3E( );
void setroute4F( );
void abnorm1( );
void abnorm2( );
void dnmainbklock( );
void dnmainapplock( );
void mlthrecep( );
char route[5]="*1A*";
int x1=strcmp(route1, route);
int x2=strcmp(route2, route);
int x3=strcmp(route3, route);
int x4=strcmp(route4, route);
int x5=strcmp(route5, route);
int x6=strcmp(routechk1, route);
int x7=strcmp(routechk2, route);
int x8=strcmp (route1Abklock, route);
int x9=strcmp(route1Aapplock, route):
int x10=strcmp(route1Amlthrecep, route];
switch(x1)
case 0:
if(x1==0)
{
setroute1A( );
}
switch(x2)
        case 0:
if(x2==0)
{
        setroute1B( );
}
switch(x3)
    case 0:
if(x3==0)
{
     setroute2E( );
}
```

```
switch(x4)
      case 0:
if(x4= =0)
{
    setroute3E( );
}
switch(x5)
      case 0:
if(x5==0)
{
    setroute4F( );
}
switch(x6)
case 0:
if(x6==0)
        {
            abnorm1( );
        }
switch(x7)
      case 0:
        if(x7==0)
        {
            abnorm2( );
        }
switch(x8)
      case 0:
        if(x8==0)
          {
                dnmainbklock( );
          }
switch(x9)
      case 0:
        if(x9= =0)
          {
                dnmainapplock( );
          }
switch (x10)
      case 0:
        if (x10= =0)
          {
                mlthrecep( );
          }
getch ( );
return;
}
```

Fig. 7.9 An example of a 'Switch…case' function

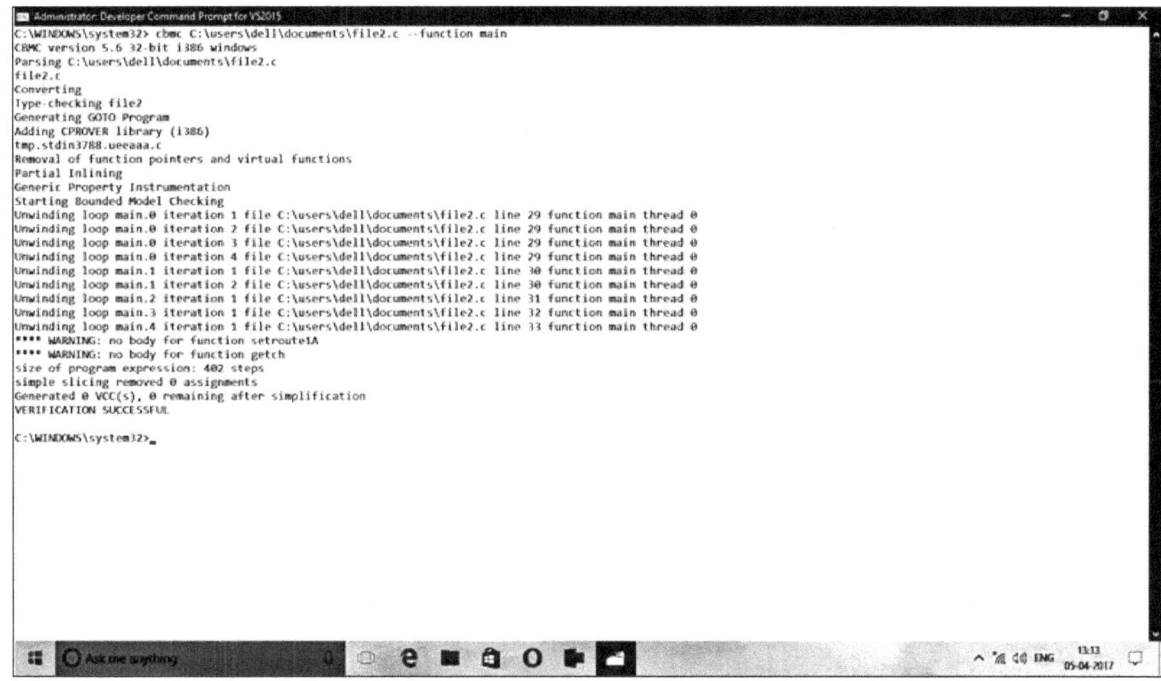

Fig. 7.10 Screenshot of verification of 'Switch……case' function

References

[1] CENELEC Standard EN-50128 – software for railway control and protection system

[2] CENELEC Standard EN-50129 – safety related electronic systems for signalling

[3] E. Clarke and D. Kroening "ANSI-C Bounded Model Checker User Manual" – Carnegie Mellon University Aug2 2006.

[4] W.J. Cullyer, S.J.Goodenough and B.A. Wichmann "The choice of computer languages for use in safety-critical systems" – Software Engineering Journal March 1991

[5] "Guide to Software Verification and Validation" – European Space Agency Board for Software Standardisation and Control – March 1995.

[6] A. Gurfinkel "Introduction to CBMC"– Carnegie Mellon University Nov19 2012.

[7] "IEEE Guide for Software Verification and Validation Plans" – Software Engineering Standards Committee of the IEEE Computer Society – Dec 2 1993.

[8] Indian Railway specification no. RDSO/SPN/144/2014-Safety and reliability requirement of electronic signalling equipment-www.rdso.indianrailways.gov.in

[9] Indian Railway specification no. RDSO/SPN/192/2005 – Electronic Interlocking-www.rdso.indianrailways.gov.in

[10] D. Kroening et al "CBMC-C Bounded Model Checker" – Carnegie Mellon University Jan 5 2009.

[11] MISRA-C "Guidelines for the use of the C language in critical systems" issued by Motor Industry Software Reliability Association(MISRA) – U.K. October 2004.

[12] NASA - GB - 8719.13 "NASA Software Safety Guidebook" issued by National Aeronautics and Space Administration as a Technical Standard – 31st March 2004.

[13] Patrick D.T. O'Connor et al. "Practical Reliability Engineering" IV edition-John Wiley & Sons 2002

[14] C.E. Torp "Method of software validation" – published by Nordtest Finland – March 2003

## CHAPTER 8

FORMAL VERIFICATION OF RAILWAY INTERLOCKING

8.1 Need for formal verification

Formal verification methods have attained importance as it has been observed that subtle design errors not found even after months of simulation could be located by applying formal methods [4]. Formal verification now supplements traditional methods of simulation and testing. By reducing the system (both hardware and software) to a logical model with finite states, the verification can be done faster even upto 10^{100} states. Formal verification can therefore save on cost and time compared to the traditional system which may require many years to stimulate the operational states. The preferred formal verification method is the "temporal logic model checking", now widely used.

8.2 Model checking

If the system to be verified can be modelled with a finite number of states, model checking can be employed to check for system correctness. A finite description of the system is compared to a logical statement which reflects the specification (in other words, model of the specification) and if it is satisfiable, the system is considered to be correct [5]. An overview of the model checking is given in Fig 8.1. If the model checking is not satisfiable, a counter example is released, which gives an idea of the violation of a property [7]. The specification is normally expressed in temporal logic.

8.2.1 As per Baier et al. [1], "Model checking is an automated technique that, given a finite state model of a system and a formal property, systematically checks whether this property holds for (a given state in) that model".

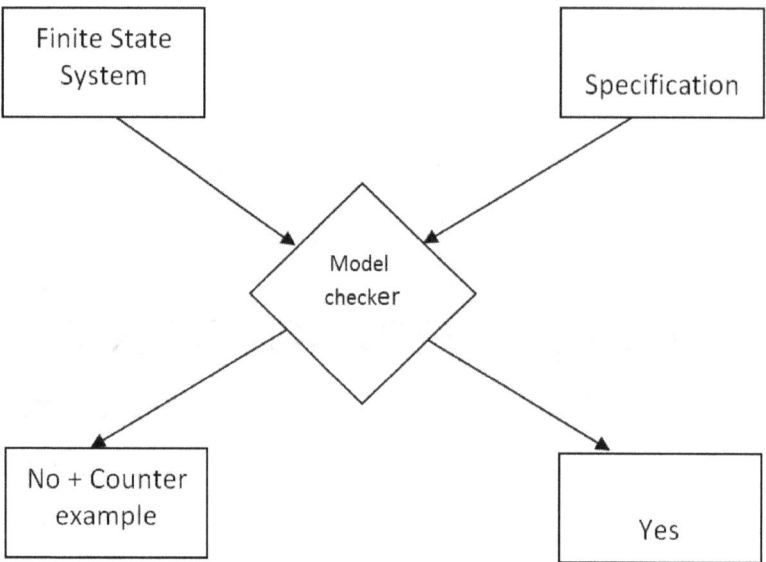

Figure 8.1 overview of model checking

8.3 Temporal logics

Temporal logics are widely used specification languages in model checking for specifying the properties of finite state systems, among many languages such as B, Z, SDL (Specification and Description language) etc. Of the two temporal logics, LTL (Linear Temporal Logic) and CTL (Computation Tree Logic), CTL is more advantageous as 'branching' can be depicted.

8.3.1 A simple example of temporal logic

Consider the simple program below, taken from [6]

```
    byte x;
1: x = 13;
x = x/2 + 1
2: if x < = 0
    goto 1;
else
    x = x/2;
    goto 2
```

This program can be represented as a system with four states as follows:

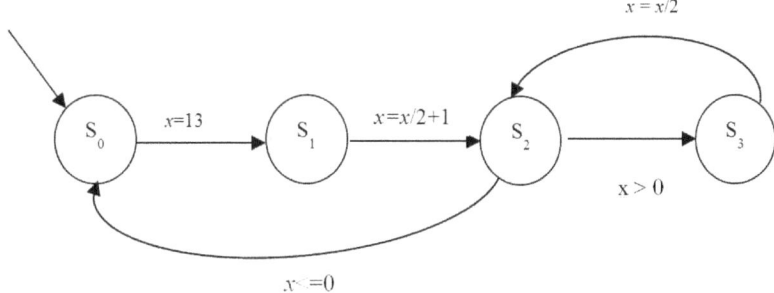

Figure 8.2 Transition diagram of the program above

8.3.2 For example, let us define two properties p and q and see whether they are continuously true. *Property p: value of x is odd and property q: value of x is 13.* Assuming that x is 0 initially, the run of the program gives p as *false (S_0, 0), true (S_1, 13), true (S_2, 7), true (S_3, 7), true (S_2, 3), true (S_3, 3), true (S_2, 1), true (S_3, 1), false (S_2, 0), false (S_0, 0)* taking integers into account. If p is specified as true always, then the false states should be made unreachable. As for property q, it is true only in one transition i.e. *(S_1, 13)*, at all others, it is false. As the run of the program is infinite, it has to be made finite as per the requirement, by modifying the program or use the technique of abstraction. In model checking, the state at each transition is verified with the property specified and where the requirement is not met, a counter example is suggested in order to modify the program to suit the requirement all time.

8.4 Symbolic model checking or verification

In *Symbolic Model Verification (SMV)*, the system to be verified is expressed in the form of a finite Boolean function such as $F \equiv f (v_1, v_2, ------ v_n \lor v_1', v_2', ----- v_n')$ where $v_1, v_2 ... v_n$ are variables at the first state and $v_1', v_2', ... v_n$ are variables at the second transition. This can be converted into a CTL (Computation Tree Logic) and in the SMV (Symbolic Model Verifier), a software tool, the logic in the form of OBDD (Ordered Binary Decision Diagram) is compared with that of the specification at each transition path and incongruity is brought out.

8.4.1 Kripke Structure

The state transition diagram of the system to be verified is expressed in the form of Kripke Structure[2], a minimum of three-tuple i.e. M(S, R, L) where

S = finite set of states $\{S_1, S_2 ... S_n\}$

R =   transition relation – expressed in pairs

L = set of labels assigned to states – the properties are assigned as labels.

An infinite computation tree with the initial state as the root, is derived from the Kripke structure while adopting the Computation Tree Logic (CTL) as the temporal logic and the path quantifiers as well as temporal operators are chosen to describe the properties of the system.

The path quantifiers are

A – for all computation paths

E – for a computation path which exists

The temporal operators combined with the states are called state operators and are as below

G a : a (state) holds globally

F a : a holds eventually (or in future)

X a : a holds at the next state

a **U** b: a holds until b (state) holds

a **R** b : a releases b

Every temporal operator (**G, F, X, U, R**) must be immediately preceded by a path quantifier (**A, E**)

Common basic CTL formulae are as follows.

AG(f): true in S if f (property) holds in every state along all paths emanating from S (for all paths...globally)

EG(f): true in S if f holds in every state along some path emanating from S (there exists a path...)

AF(g): for all paths, eventually there is a state in which g holds

EF(g): there exists a path which eventually contains a state in which g is true.

An example for **AG**(g) is illustrated below:

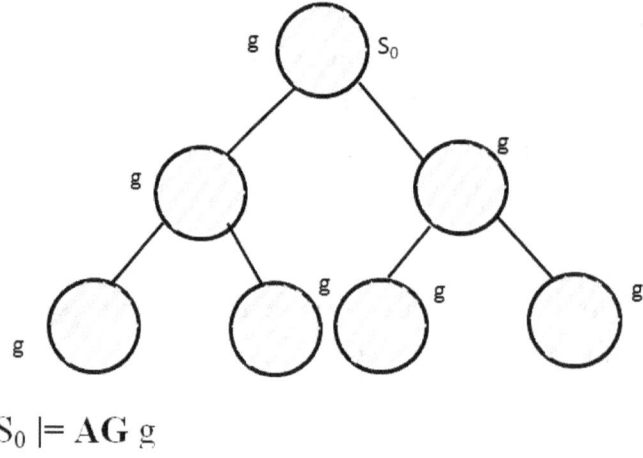

$$S_0 \models \textbf{AG } g$$

Figure 8.3

In the figure 8.3, the property g holds in every state along all paths emanating from S_0.

Similarly in LTL (Linear Temporal Logic), the prefix **A** or **E** is not required as there is a single path without branching. The equivalent semantic for global holding of property g is as below:

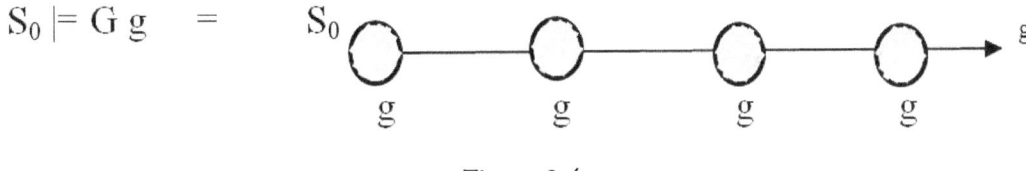

Figure 8.4

8.5 Symbolic Model Verification with NuSMV

The NuSMV software tool can be obtained from www.nusmv.fbk.eu. This tool has been developed by ITC-irst (European Centre for Information Technology) and CMU (Carnegie Mellon University) USA. The inputs to NuSMV have to be in an ordered language similar to a hardware description language.

8.5.1 The following is a simple program in the NuSMV language [3]

```
MODULE main
    VAR
        request : boolean;
        state : {ready, busy};
    ASSIGN
        init(state) := ready;
        next(state) := case
                    state = ready & request = TRUE : busy;
                    TRUE                           : {ready, busy};
                esac;
```

The state variables, request and state have to be declared. Request is declared as boolean which can be FALSE or TRUE. The variable state is a scalar variable, which has the symbolic values ready or busy. The following assignment (ASSIGN) sets the individual value of the variable state to ready, it means it can be FALSE or TRUE. The case segment sets the next value of the variable state to the value busy (after the colon) if its current value is ready and request is TRUE. Otherwise (the TRUE before the colon) the next value for state can be any in the set {ready, busy}. The variable request is not assigned, which means, it can assume any value.

8.5.2 To understand the safety property, the following semaphore program is studied [3]

```
MODULE main
        VAR
            semaphore : boolean;
                proc 1 : process user (semaphore);
                proc 2 : process user (semaphore);
            ASSIGN
                init (semaphore) : = FALSE;
        SPEC AG ! (proc1.state = critical & proc2.state = critical)
        SPEC AG  (proc1.state = entering →AF proc1.state = critical)
MODULE  user(semaphore)
        VAR
            state: { idle, entering, critical, exiting};
        ASSIGN
            init(state) := idle;
            next (state) : =
                    case
                        state = idle            :{idle, entering};
                        state = entering & ! semaphore   : critical;
                        state = critical        : {critical, exiting};
                        state = exiting         : idle;
                        TRUE                    : state;
                    esac;
            next (semaphore) : =
                    case
                        state = entering : TRUE;
                        state = exiting : FALSE
                        TRUE            :semaphore;
                    esac;
    FAIRNESS
        Running
```

8.5.3 The semaphore program is similar to a 'semaphore' in railway signalling where one operation or one train is permitted, here only one process turning critical is allowed. The following specification is included in the program i.e.

> AG ! (proc1.state = critical & proc2.state = critical)

This is in CTL as already explained and if the output on running the program says it is *true*, then it means that the two processes proc1 and proc2 can never be critical at the same time.

By running NuSMV with the command

> system_prompt > **NuSMV semaphore.smv**

the following output is obtained

--- specification AG (!proc1.state = critical & proc2.state = critical))
--- is true

which means that the two processes cannot be 'critical' at the same time

Regarding the other specification i.e.

> AG (proc1.state = entering → AF proc1.state = critical)

the output obtained is

> -- is false

which means that proc1.state cannot enter the 'critical' state as the loop is such that it tries to enter when proc2.state is turned critical which is not possible. The program gives a counter example why it happens and the program can be modified with this knowledge.

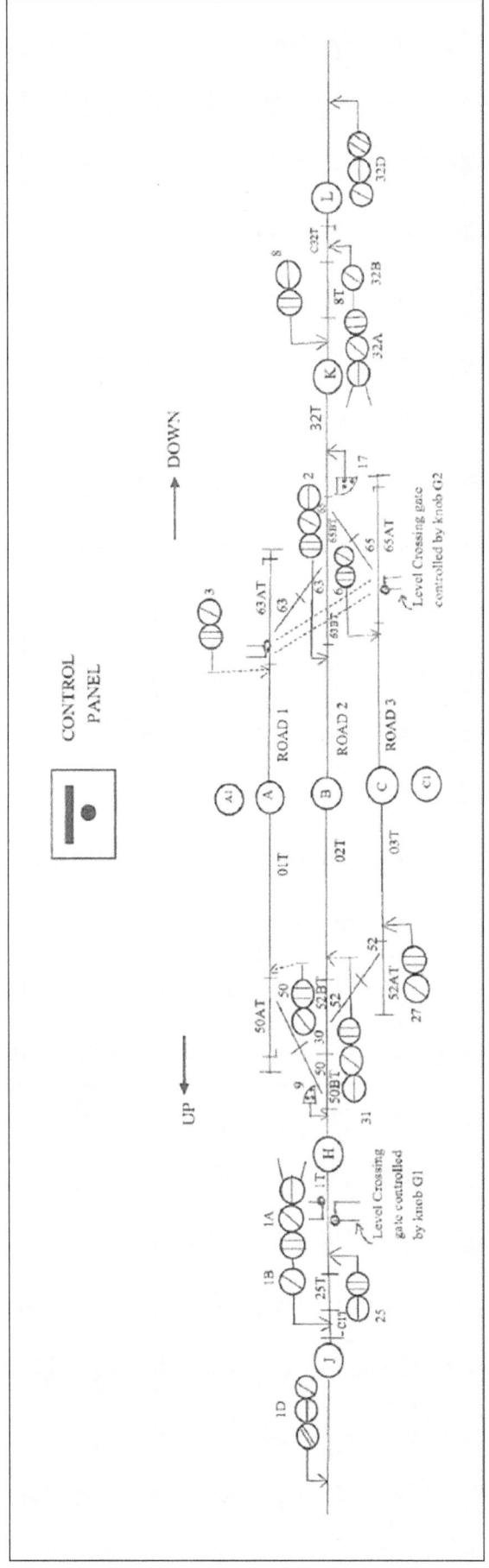

Figure 8.5 Sketch of signalling at station 'Z'

Table 8.1 Partial table of control for station 'Z'. showing down side signals

Description	Route No.	Locks & detects points	Locks routes	Controlled by tracks	App. tracks	Back lock tracks	Remarks
Down home signal to road 1 with 63 reverse	1A-A	50R, 63R, 65N	1B-A, 9A, 17A, 25J, 30H, 32A-A, 32B-A	25T, 1T, 50BT, 50AT, 01T, 63AT, 63BT, 65BT, 32T	120 sec	25T, 1T, 50BT, 50AT	UG* on, gate keys and shunting keys in
Down home signal to road 1 with 63 normal	1A-A$_1$	50R, 63N	1B-A, 9A, 17B, 17C, 25J, 30H, 32B-B, 32B-C	25T, 1T, 50BT, 50AT, 01T, 63AT	120 sec	25T, 1T, 50BT, 50AT	UG* on, gate keys and shunting keys in
Down home signal to road 2	1A-B	50N, 52N, 63N, 65N	1B-B, 9B, 17B, 25J, 31H, 32A-B, 32B-B	1T, 50BT, 52BT, 02T, 63BT, 65BT, 32T, 25T	120 sec	1T, 50BT, 52BT, 25T	gate keys and shunting keys in
Down home signal to road 3 with 65 reverse	1A-C	50N, 52R, 65R	1B-C, 9C, 17C, 25J, 27H, 32A-C, 32B-C	1T, 50BT, 52BT, 52AT, 03T, 65AT, 65BT, 32T, 25T	120 sec	25T, 1T, 50BT, 52BT, 52AT	UG* on, gate keys and shunting keys in
Down home signal to road 3 with 65 normal	1A-C$_1$	50N, 52R, 65N	1B-C, 9C, 17B, 25J, 27H, 32B-B	25T, 1T, 50BT, 52BT, 52AT, 03T, 65AT	120 sec	25T, 1T, 50BT, 52BT, 52AT	UG* on, gate keys and shunting keys in
Calling on below down home signal to road 1	1B-A	50R	1A-A, 1A-A$_1$, 3K, 6K, 9A, 17A, 17B, 17C, 25J, 30H, 32A-A, 32A-C, 32B-B, 32B-C	–	120 sec	25T, 1T, 50BT, 50AT	C1T occupied, gate key and shunting key down side in
Calling on below down home signal to road 2	1B-B	50N, 52N	1A-B, 2K, 3K, 6K, 9B, 17A, 17B, 17C, 25J, 31H, 32A-A, 32A-B, 32A-C, 32B-A, 32B-B, 32B-C	–	120 sec	25T, 1T, 50BT, 52BT	C1T occupied, gate key and shunting key down side in
Calling on below down home signal to road 3	1B-C	50N, 52R	1A-C, 1A-C$_1$, 6K, 9C, 17B, 17C, 25J, 27H, 32A-C, 32B-B, 32B-C	–	120 sec	1T, 50BT, 52BT, 52AT, 25T	C1T occupied, gate key and shunting key down side in
Down starter from road 2	2-K	63N, 65N	1B-B, 9B, 17B, 32A-B, 32B-B	63BT, 65BT, 32T	02T (120 sec)	63BT, 65BT	gate key and shunting key up side in
Down starter from road 1	3-K	63R, 65N	1B-A, 1B-B, 9A, 9B, 17A, 32A-A, 32A-A$_1$, 32B-A	63AT, 63BT, 65BT, 32T	01T (60 sec)	63AT, 63BT, 65BT	gate key and shunting key up side in
Down starter from road 3	6-K	65R	1B-A, 1B-B, 1B-C, 9A, 9B, 9C, 17C, 32A-C, 32A-C, 32B-C	65AT, 65BT, 32T	03T (60sec)	65AT, 65BT	gate key and shunting key up side in
Down advanced starter	8-L	–	17A, 17B, 17C, 32A-A, 32A-A$_1$, 32A-B, 32A-C, 32A-C$_1$, 32B-A, 32B-B, 32B-C	8T	–	–	controlled by token less block instrument
Down shunt to clear of up starter 30 (road 1)	9-A	50R	1A-A, 1A-A$_1$, 1B-A, 3K, 6K, 17A, 25J, 30H, 32A-A, 32A-C, 32B-A, 32B-B, 32B-C	50BT, 50AT	1T (60 sec)	50BT, 50AT	gate key, shunting key in
Down shunt to clear of up starter 31 (road 2)	9-B	50N, 52N	1A-B, 1B-B, 2K, 3K, 17B, 25J, 31H, 32A-A$_1$, 32A-B, 32A-C$_1$, 32B-A, 32B-B, 32B-C	50BT, 52BT	1T (60 sec)	50BT, 52BT	gate key, shunting key in
Down shunt to clear of up starter 27 (road 3)	9-C	50N, 52R	1A-C, 1A-C$_1$, 1B-C, 6K, 17C, 25J, 27H, 32A-C, 32B-B, 32B-C	50BT, 52BT, 52AT	1T (60 sec)	50BT, 52BT, 52AT	gate key, shunting key in

*UG – Route indicator with yellow lights

8.6 Generation and Verification of Control Table for signalling at a station applying SMV

The signalling at station 'Z' (Fig 8.5) has been taken to generate a control table which has been verified according to the signalling principles. The signalling is based on that at 'Kumaramangalam' on the Trichy-Karaikudi single line section of Southern Railway, India. This is chosen because of complex interlocking involving main signals, calling-on signals and shunt signals. Approach locking and back locking has been omitted as SMV tool does not support it. Partial control table showing down side signals only is depicted in Table 8.1. The control table generated is found to be an exact replica of the table made by the drawing office of the railway. The various stages for generation and verification of the control table are illustrated in Fig. 8.6.

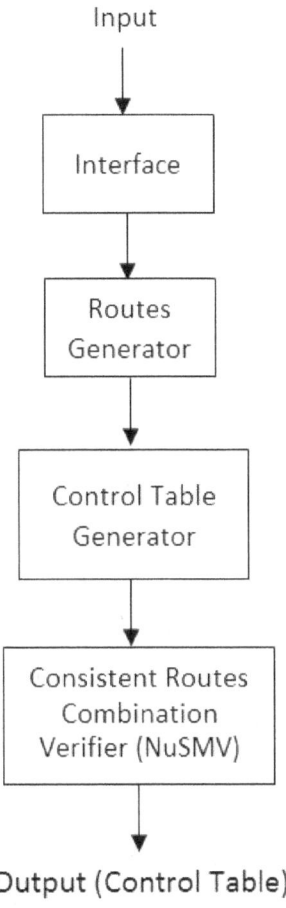

Figure 8.6 Generation and verification of control table for station 'Z'

8.6.1 The input in the form of a specified format having 6 sections is generated in the interface taking signalling data marked on the signalling plan converted to a graph. The first section consists of track circuits of the whole station sorted in rows, each row containing two track circuits, one track circuit common to both lines acting as a fulcrum to provide continuity to the graph. The second section lists the conflicting track circuits at each of the four cross-over points, which a train cannot traverse. The third section lists the signals and the track circuits near which the signals are located. The fourth section lists the knobs (e.g. A, B, A_1 etc.) and the track circuits on which they are located. The fifth section groups the track circuits for the three reception lines.

The sixth section lists all the routes for the station. Each section is separated by 'END'. To form this text input, the signalling plan is converted to a graph with different types of vertices and edges, representing the signal switches, track circuits, points and exit knobs. The input file (text format) is given in Fig 8.7.

8.6.2 The *routes generator* takes the formatted input and using a slightly modified version of Depth First Search algorithm in Ruby language brings out the routes, clubbing the signal switches, knobs, points and controlling track circuits. The *Control Table Generator* module also uses Ruby programming language on Ubuntu 14.04 LTS OS. It checks the status of points for each route and feeds the rudimentary control table to the *Consistent Routes Combination Verifier* which also consists of NuSMV model checker. The model checker verifies that trains following different routes do not collide, also two routes which have common points on their paths are not enabled at the same time. In the NuSMV program, the routes are in the form of modules and they are allowed to move as synchronous non-deterministic transition model from the main module. The condition to be checked for non-collision is given in the form of an LTL (Linear Temporal Logic) formula depicting two trains following different routes, as below.

G! (train_1.track id = train_2.track id)

With the formula above, conflicting routes can be identified and a valid control table is generated at the output. In the experiment carried out and described in [8], the control table generated was found to be an exact replica of the original table formulated in the drawing office of the controlling railway (Southern Railway). The experiment was carried out by a graduate student (K. Sriram) under the guidance of S. Sheerazuddin, Assoc. Professor of Computer Science at SSN College of Engineering located on the outskirts of Chennai.

8.6.3 A copy of the Ruby language program for *control table generator* is attached at Appendix D. This is one of sixteen Ruby language programs for the various functions for the generation of control table. The rudimentary control table so generated is verified and modified by means of the SMV language inputs which provide for the elimination of conflicts. Each pair of conflicting routes over the common points is assigned a module. An SMV program for the conflicting pair of routes 1A_A and 32A_A is illustrated in Fig 8.8. Route 1A_A has six conflicting routes, similarly for other routes. For all the routes in the signalling plan for station 'Z' there are 202 SMV files which are referred to by the *Consistent Routes Combination Verifier* program in Ruby, before generating the final control table.

8.6.4 The control table is generated in four parts and the first part is illustrated in Table 8.2. It is seen that the table is identical to the original table, omitting the approach locking and back locking, the inclusion of which require further development.

8.7 Application of SMV for the Verification of Control Tables

The verification of the control table of station 'Z' by SMV is a work in progress as the layout is complicated with a variety of signals i.e. main signals, calling-on signals and shunt signals. Also many problems, contradictions and constraints have come to the fore, which require detailed examination and solution.

8.7.1 A simpler layout of station 'Y' is taken for verification of the control table. The signalling plan and interlocking table for the down side of the station 'Y' are already shown in Fig 7.2. An SMV program for verification of 'no collision' and 'no derailment' is reproduced in Fig 8.9. The program is executed in batch mode, through 'NuSMV-2.6.0-win64' software tool downloaded from the website http://nusmv.fbk.eu. A small train (train 1) is stabled permanently on down main line occupying track circuit 01AT. The train 2 which is moving or supposed to be in the rear of signal '1' cannot move and collide with train 1 as the route '1A' cannot be set and the signal 1 cannot be 'off'. This is confirmed by the affirmation of the specification as seen in the screenshot of the output shown in Fig. 8.10. Similarly 'no derailment' can be confirmed indirectly by showing that the points 51 and 53 can be changed only when the route '1A' is not set or the signal 1 is not 'on'. This is confirmed in the output. The two specifications listed below have been confirmed as true in the output of the NuSMV as seen in the Fig. 8.10.

```
AG(train1on _01AT = TRUE & train2on _01AT=FALSE)
 AF(points_51=FALSE & points_53=FALSE& route_1A_set=FALSE)
```

```
C1T 25T
25T 1T
1T 50BT
50BT 50AT
50BT 52BT
52BT 52AT
52BT 02T
02T 63BT
50AT 01T
01T 63AT
63AT 63BT
63BT 65BT
52AT 03T
03T 65AT
65AT 65BT
65BT 32T
32T 8T
8T C32T
END
52BT 50BT 50AT 50
02T 52BT 52AT 52
```

02T 63BT 63AT 63

63BT 65BT 65AT 65

END

C1T (1B down calling_on_home)

C1T (1A down home)

1T (25 up advanced_starter)

1T (9 down shunt)

01T (30 up starter)

02T (31 up starter)

03T (27 up starter)

01T (3 down starter)

02T (2 down starter)

03T (6 down starter)

32T (17 up shunt)

32T (8 down advanced_starter)

C32T (32A up home)

C32T (32B up calling_on_home)

END

C1T J

1T H

01T A A1-P

02T B

03T C C1-P

32T K

C32T L

END

T1 C1T 25T 1T 50BT 52BT 02T 63BT 65BT 32T 8T C32T

T2 50AT 01T 63AT

T3 52AT 03T 65AT

END

1A A A1 B C C1

1B A B C

2 K

3 K

6 K

8 L

9 A B C

17 A B C

25 J

27 H

30 H

31 H

32A A A1 B C C1

32B A B C

Figure 8.7 Input File – Station 'Z'

```
-- *****
-- EVERY ROUTE IS A MODULE. A TRAIN WILL MOVE IN A ROUTE.
-- *****
MODULE route_1A_A()
 VAR
  track_id : {"25T", "1T", "50BT", "50AT", "01T", "63AT", "63BT", "65BT", "32T"};
 ASSIGN
  init(track_id) := "25T";
  next(track_id) := case
              track_id = "25T" : {"25T", "1T"};
              track_id = "1T" : {"1T", "50BT"};
              track_id = "50BT" : {"50BT", "50AT"};
              track_id = "50AT" : {"50AT", "01T"};
              track_id = "01T" : {"01T", "63AT"};
              track_id = "63AT" : {"63AT", "63BT"};
              track_id = "63BT" : {"63BT", "65BT"};
              track_id = "65BT" : {"65BT", "32T"};
              TRUE : track_id;
            esac;
MODULE route_32A_A()
 VAR
  track_id : {"8T", "32T", "65BT", "63BT", "63AT", "01T", "50AT", "50BT", "1T"};
 ASSIGN
  init(track_id) := "8T";
  next(track_id) := case
              track_id = "8T" : {"8T", "32T"};
              track_id = "32T" : {"32T", "65BT"};
              track_id = "65BT" : {"65BT", "63BT"};
              track_id = "63BT" : {"63BT", "63AT"};
              track_id = "63AT" : {"63AT", "01T"};
              track_id = "01T" : {"01T", "50AT"};
              track_id = "50AT" : {"50AT", "50BT"};
              track_id = "50BT" : {"50BT", "1T"};
              TRUE : track_id;
            esac;
MODULE main()
 VAR
  train_1A_A : route_1A_A();
  train_32A_A : route_32A_A();
LTLSPEC
 G !(train_1A_A.track_id = train_32A_A.track_id);
```

Figure 8.8 Conflicting routes 1A_A and 32A_A–SMV language program

Table 8.2 Control Table Generated for Station Yard 'Z' (1/4)

Description	Signal	Label	Locks and Detects Points	Locks Routes	Controlled by Tracks
Down from Home signal	1A	A	50-R, 63-R, 65-N	1B-A, 9-A, 17-A, 25-J, 30-H, 32A-A, 32B-A	25T, 1T, 50BT, 50AT, 01T, 63AT, 63BT, 65BT, 32T
Down from Home Signal	1A	A1	50-R, 63-N	1B-A, 9-A, 17-B, 17-C, 25-J, 30-H, 32B-B, 32B-C	25T, 1T, 50BT, 50AT, 01T, 63AT
Down from Home Signal	1A	B	50-N, 52-N, 63-N, 65-N	1B-B, 9-B, 17-B, 25-J, 31-H, 32A-B, 32B-B	25T, 1T, 50BT, 52BT, 02T, 63BT, 65BT, 32T
Down from Home Signal	1A	C	50-N, 52-R, 65-N	1B-C, 9-C, 17-C, 25-J, 27-H, 32A-C, 32B-C	25T, 1T, 50BT, 52BT, 52AT, 03T, 65AT, 65BT, 32T
Down from Home Signal	1A	C1	50-N, 52-R, 65-N	1B-C, 9-C, 17-B, 25-J, 27-H, 32B-B	25T, 1T, 50BT, 52BT, 52AT, 03T, 65AT
Down from calling on signal	1B	A	50-R	1A-A, 1A-A1, 3-K, 6-K, 9-A, 17-A, 17-B, 17-C, 25-J, 30-H, 32A-A, 32A-C1, 32B-A, 32B-B, 32B-C	–
Down from calling on signal	1B	B	50-N, 52-N	1A-B, 2-K, 3-K, 6-K, 9-B, 17-A, 17-B, 17-C, 25-J, 31-H, 32A-A1, 32A-B, 32A-C1, 32B-A, 32B-B, 32B-C	–
Down from calling on signal	1B	C	50-N, 52-R	1A-C, 1A-C1, 6-K, 9-C, 17-B, 17-C, 25-J, 27-H, 32A-C, 32B-B, 32B-C	–

```
MODULE main
    VAR
        points_51:boolean;
        points_53:boolean;
        route_1A_set:boolean;
        route_1B_set:boolean;
        track_1T:boolean;
        track_1AT:boolean;
        track_51T:boolean;
        track_01T:boolean;
        track_01AT:boolean;
        track_53T:boolean;
        DRP:boolean;
        train1on_01AT:boolean;
        train2_move:boolean;
        train2on_01AT:boolean;
ASSIGN
        init(track_1T):= TRUE;
        init(track_1AT):= TRUE;
        init(track_51T):= TRUE;
        init(track_01T):= TRUE;
        init(track_01AT):=FALSE;
        init(track_53T):=TRUE;
        init(points_51):=TRUE;
        init(points_53):= TRUE;
        init(DRP):=TRUE;
        init(route_1A_set):= FALSE;
        init(route_1B_set):= FALSE;
        next(track_1T):= track_1T;
        next(track_1AT):= track_1AT;
        next(track_51T):= track_51T;
        next(track_01T):= track_01T;
        next(track_01AT):= track_01AT;
        next(track_53T):= track_53T;
        next(points_51):= case
          route_1A_set:TRUE;
          !route_1A_set:FALSE;
          TRUE:points_51;
            esac;
        next(points_53):= case
          route_1A_set:TRUE;
          !route_1A_set:FALSE;
          TRUE:points_53;
              esac;
        next(DRP):= DRP;
        next(route_1A_set):= case
```

```
    track_1T & track_1AT & track_51T & track_01T &
    track_01AT& track_53T & points_51 & points_53&!route_1B_set: TRUE;
    !track_1T | !track_1AT | !track_51T | !track_01T | !track_01AT |!track_53T | !points_51 |
    !points_53|route_1B_set:FALSE;
    TRUE:route_1A_set;
        esac;
next(route_1B_set):= case
    !route_1A_set & track_1T & track_1AT & track_51T & track_53T & !points_51 & !points_53 & DRP:TRUE;
    route_1A_set:FALSE;
    TRUE: route_1B_set;
        esac;
init(train1on_01AT):=TRUE;
next(train1on_01AT):=train1on_01AT;
init(train2_move):=TRUE;
next(train2_move):=case
    !route_1A_set:TRUE;
    route_1A_set:FALSE;
    TRUE:train2_move;
        esac;
init(train2on_01AT):= FALSE;
next(train2on_01AT):= case
    route_1A_set:TRUE;
    !route_1A_set |!points_51 |!points_53: FALSE;
    TRUE:train2on_01AT;
        esac;
--No collision
SPEC
 AG (train1on_01AT=TRUE & train2on_01AT=FALSE)
--No possibility of derailment
SPEC
 AF (points_51=FALSE & points_53=FALSE&route_1A_set=FALSE)
```

Figure 8.9 SMV program for station 'Y'

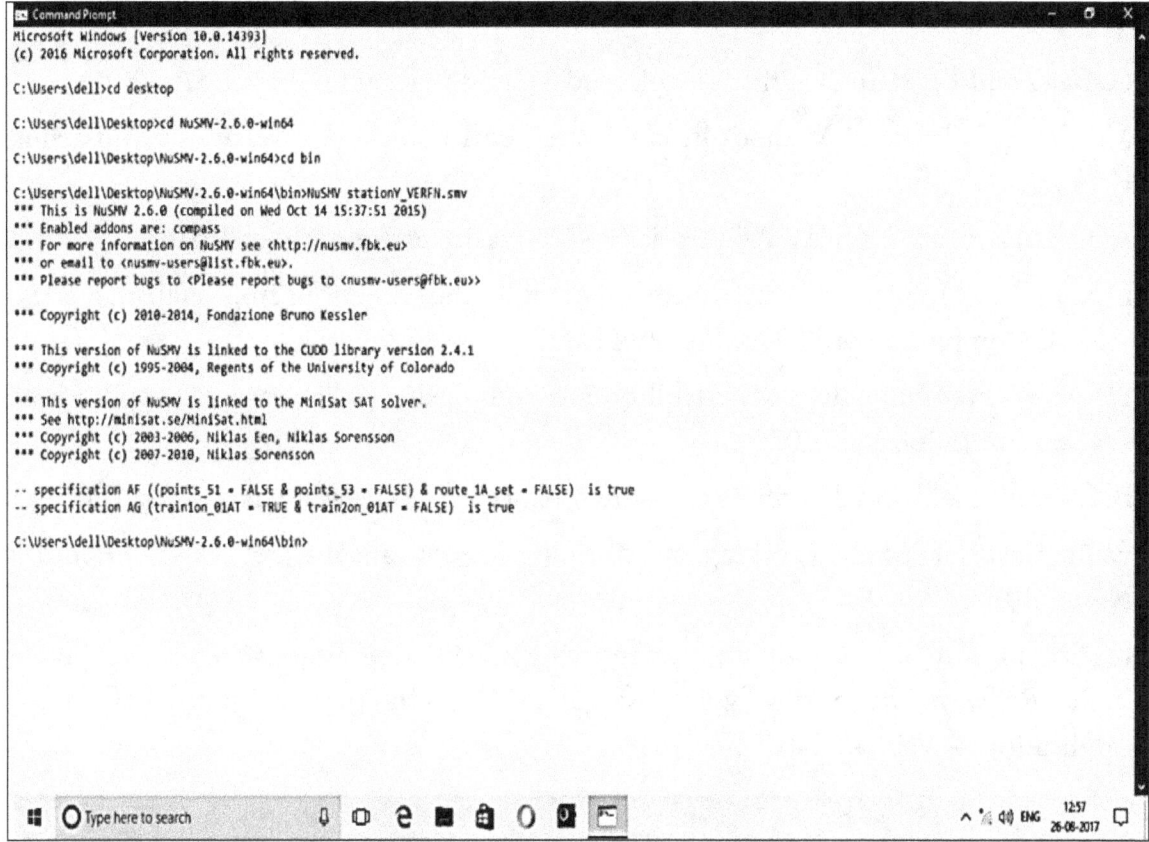

Figure 8.10 Screenshot of the output of NuSMV – station 'Y'

This indirect confirmation had to be resorted to as the symbolic enumeration with scalar variables and the following variable assignment gave negative results and the simulation as follows,

train1.pos : {moving, 01AT};

train2.pos: {moving. 01AT};

moved over to unwanted states, though not permitted in the coding. The train positions had to be assigned boolean values as declared below,

train1on_01AT : boolean;

train2_move : boolean;

train2on_01AT : boolean;

also shown in the SMV coding given in Fig. 8.9. This gave precise results.

References

[1] C. Baier and J.P. Katoen "Principles of Model Checking", MIT Press USA 2008.

[2] M. Ciesielski "Formal Methods in Hardware Verification" – University of Massachusetts USA 2001.

[3] A. Cimatti et. al. "NuSMV 2.6 tutorial", – University of Trento Italy 2010.

[4] D.L. Dill and J. Rushby, "Acceptance of Formal Methods: Lessons from Hardware Design" – IEEE Computer, April 1996, Vol. 29 No. 4.

[5] Elaine Rich, "Automata, computability and complexity – Theory and Applications" – Pearson Education Inc 2008.

[6] B. Meenakshi, "Formal Verification" – 'Resonance' of Honeywell Labs – May 2005.

[7] Paulo Traverso, "Model checking and planning for critical software" – IRST Trento Italy Nov. 7 2002.

[8] K. Sriram and S. Sheerazuddin, "Identifying Conflicting Routes in Control Table of Indian Railways Interlocking System using NuSMV" – International Journal of Computer Applications – Vol. 146 No.7-July 2016.

## CHAPTER 9

APPLICATION OF SYSTEMS ENGINEERING

9.1 Indian Context

While the full scale application of Systems Engineering is not necessary in the Indian context, for finalisation of *Requirements*, *Design*, *Manufacture* and *installation*, there is need to consider application in *Operation* and *Maintenance* of signalling systems specially as a particular application called SUBSAFE[4] was found to be highly beneficial in preventing major accidents to nuclear submarines of the U.S. Navy, since 1963 onwards. (SUBSAFE is not an acronym but a term used to signify the submarine safety program)

9.2 Systems Engineering (SE)

Systems Engineering (SE) is an interdisciplinary approach for the realization of successful systems, by developing a final product that meets the customer needs, goals and objectives. *Requirements Engineering* is the branch of systems engineering which addresses the process of identifying and then monitoring the stakeholders' needs and required functionalities [3].

9.2.1 Requirements play a key role during the whole system development life cycle and management of these requirements ensures the preservation of the information integrity during the whole system life cycle taking care of changes in the system and its environment. While considering safety, the properties of the whole system which consists of products, processes, inter-operational services etc. and above all people have to be considered.

9.2.2 Systems development life-cycle

To consider the system as a whole, it is necessary to view the V-model of the system development life cycle which is similar to the one drawn in Fig. 4.4 with variations wherever required. The system life cycle can also be viewed as a top-down implementation as shown in Fig. 9.1.

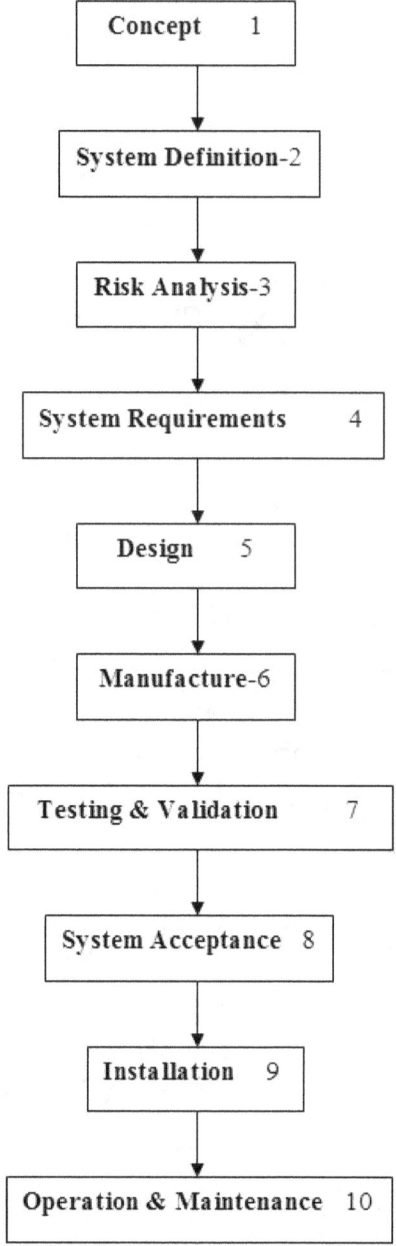

Figure 9.1 System Lifecycle

In the Indian context, for E.I. systems, the stages of Concept, System Definition and Risk Analysis (1, 2, 3) are already established and the System Requirements (stage 4) are those in the table of control for each interlocked station yard. This table is converted to equivalent relay circuits, then expressed in boolean logic, which is compiled into an executable code. This is fed into the non-erasable memory, as the designed and manufactured system is already existing. For the original system, the stages of 1, 2, 3, 4 and 5 (Fig. 9.1) are carried out by the E.I. system supplier and the stage of manufacture (stage 6) is just fabrication with selected OTS (Off The Shelf) components. The testing, validation and acceptance is carried out by R.D.S.O as detailed in paragraph 4.8.9 and sub-paragraphs. The installation and commissioning of the E.I system is carried out jointly by the supplier and the local railway through specialised staff. This also involves rigorous functional testing at site in addition to testing and validation (stage 7) at the supplier's site. The operation and maintenance is carried out by the local railway staff as per the Signal Engineering Manual (in vogue) and the Station Working Rules which include the formal General Rules of the Indian Railways.

9.3 Application of Systems Engineering to Railway Projects

As per Bruce Elliott [2], Railway Projects should be prepared to adapt conventional SE (System Engineering) approaches *but only where* "there is good reason to deviate from established proven methods." In India, projects in which copper cables and unmanned equipments need to be installed along the track, have been failures due to frequent thefts which are uncontrollable even with the maximum security owing to the fact that police protection 24 × 7 is just not possible. Now with the advent of optic fibre cable, cable thefts could be avoided but costly unmanned equipments along the tracks faraway from the station yards are still to be avoided. For example, projects of Central Traffic Control on two zonal railways have been failures, also automatic closure of level crossing gates by approaching trains were also beset with problems of thefts. Projects in urban areas such as Automatic Signalling have survived due to less frequent thefts with frequent train movements and movement of civilian population alongside the tracks. A viable alternative is transmission of information by wireless and concentration of equipment in the locomotive or guard's van, such as cab signalling without any wayside equipment. This is likely to be planned for high speed trains. At present, signalling installations are found only in yards both large and small. Sophisticated projects much as CTM (Central Traffic Management) through GSM-R, ATO (Automatic Train Operation), Cab Signalling with or without wayside signals with GSM-R etc. are not existing at present nor planned in the near future for the present speeds, however, such systems will be planned for the high speed corridors. Systems Engineering (SE) can now be applied only to new line projects such as the Konkan Railway Corporation project which however is already completed. The justification for SE is examined in the next paragraph.

9.4 Is SE justified for new line projects in Indian conditions?

A major portion of the work for a new line is that of laying a track and construction of buildings. This is done by the (Civil) Engineering department and an engineer of this department is appointed in the highest rank possible. Engineers of other departments i.e Signalling & Telecom, Traction, Electrical, Personal, Accounts etc. are of lower rank. The periodical coordination is done by the top engineer of the Civil Engineering department and more frequent coordination such as weekly, monthly is done by lower ranked engineers of this department. The issues for coordination are clear to each departmental officer and there is no need for a Systems Engineering department to work out the interfaces between the subsystems which are few and not so sophisticated i.e. there are no subsystems involving complicated hardware and software. The contractors involved in installing the various systems are also clear about the requirements for interfacing and the formation of a separate department is considered not justified, also uneconomical. Regarding the need for completing the project in quick time and with the least cost, the *critical path method* in management can be employed. This method is explained briefly below.

9.4.1 Critical path method

A model of the project is first constructed listing all activities required to complete the project, the time duration for each activity to be completed with the dependencies between the activities. The longest path of planned activities to the end of the project is calculated, as well as the earliest and latest that each activity can start and finish without making the project longer. The 'critical' activities on the longest path are determined. The planned critical path of the project can be shortened by pruning critical path activities, performing more activities in parallel, shortening the durations of critical path activities by adding resources[5]. The resources depend on the three 'M's i.e Men, Material and Methods. Preparation of a Gantt chart for the activities will help in arriving at the shortest time in which a project can be completed.

9.4.2 CPM(Critical Path Method) had been applied by me when I was in charge of re-signalling projects for yard remodelling, for replacement of overaged signalling, for doubling lines, re-signalling for 25 KV electric traction, signalling for new lines, re-signalling for gauge conversion (metre gauge to broad gauge) etc. and had helped in concentrating resources for early completion with the least cost. The most common delay in the case of re-signalling projects was the late construction of the relay rooms by the civil engineering department due to varied problems with the petty building contractors. Expecting the delays, no resources were allotted for such places of work and signalling works were commenced only when relay rooms were ready.

9.5 Need for Systems Theory

Many of the complex engineered systems post World War II are too complex for complete analysis and too organized for statistics as per Prof. N.G. Leveson [4] and fall under *organized complexity*. As complex software is inbuilt in these systems, it is difficult to apply analysis and statistics. The systems approach focuses on systems taken as a whole, not on the parts. Some properties of the systems can be treated adequately only in their entirety taking into account all facets including human and technical. The system properties depend on how the parts interact with each other and together. Analysis and design of the whole system is necessary in addition to the study of components where 'organized complexity' is observed.

9.5.1 Model of an electronic signalling system

The electronic signalling system is a closed control system with feedback by means of sensors. The model of the system is illustrated in Fig. 9.2. The diagram is self-explanatory for a signal engineer. The operator or the station master operates the switches or controls on the control panel and the computer translates these actions and conveys these commands to the actuators, in this case, the point switches and the signal driver relays etc. The feedback is received by the computer through the point detection relays and the signal lamp proving circuits etc. The displays of the positions of points and signals are exhibited on the control panel for the operator to see and take further action.

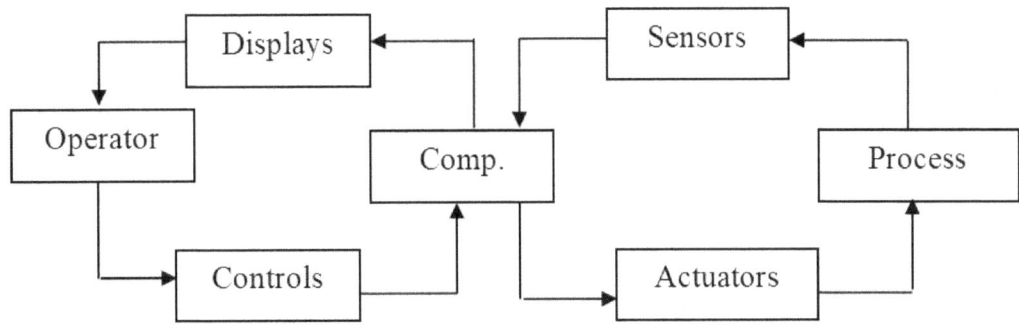

Figure 9.2 A control model of an electronic signalling system

9.6 Relationship between Systems Engineering and Safety

As per Systems theory, accidents are considered to arise from the interactions among system components and not just one factor. The operation or mal operation of the whole system is taken into account while considering the cause of the accident. Safety is affected when the system components interact with the environment. Safely is enforced by a set of constraints related to the behaviour of the system components. Accidents result only when the interactions among components violate these constraints[4]. Component failures also have to be taken into account.

9.6.1 From a study of the accidents that happened to the *Challenger* and *Columbia* space crafts by NASA of USA, it was found that systems theory provides a much better foundation for safety engineering than the classic analytic reduction approach adopted for event based models of accidents[4]. Systems Engineering along with its sub-discipline-System Safety were developed to tackle the problems and deficiencies in design, operations and management which led to accidents in aerospace.

9.6.2 Safety requirements and the introduction of safety constraints have to be considered at each stage from *Concept* to *Installation* through *Design, Manufacture, Validation* etc. in the system life cycle shown in Fig. 9.1. Alternative designs can be brought out at the end of design process and each design is assessed in terms of meeting the functional needs and financial constraints. The feasible designs are then analyzed for possible hazards and safety constraints are introduced at the subsystem level, at the interfaces and for the whole system. The best possible design which meets the performance and safety requirements is selected.

9.7 Systems-Theoretic Accident Model and Process (STAMP)

The STAMP model as annunciated by Prof. N.G. Leveson [4] as applicable to an electronic interlocking and signalling system is considered in this chapter. This model is already illustrated in Fig. 9.2 and the safety constraints required for enhancing the safety are considered. The electronic signalling system is a partially dynamic system as compared to fully dynamic systems such as nuclear reactor control or chemical process control. The operations are limited to signalling a train into the yard and dispatch of it, which are intermittent and not continuous.

9.7.1 Applying the model for enhancing safety

The use of the STAMP model not only helps in identifying the factors responsible for the accident but also leads to understanding the relationships among these. This model enables the examination of each part of the system as to how it contributed to the accident and this will provide better knowledge in engineering safer systems including the technical, managerial, organizational and regulatory aspects. The existing hazard analysis and prevention techniques are inadequate in finding the factors in the system which cause accidents and in this model the safety constraints which are necessary to prevent accidents are to be listed and the enforcement of these is to be observed. If necessary, the system has to be redesigned to enforce the constraints. Performance can also be improved by following the procedures for enforcement of these safety constraints.

9.7.2 Blaming and punishing an employee for an accident has become counterproductive, observing the history of accidents over a longer period. Taking the example of the accident in Japan (Fukuchiyama) in 2005 where a train derailed when the train driver was on the phone trying to ensure that he would not be reported for a minor infraction. While he was speaking, he could not slow down for a curve, resulting in the deaths of 106 passengers including the driver and injuries to 562 passengers. Blame and punishment had made the driver become stressful and to perform below par. Such punishments have become very common on Indian Railways and the system model which is being studied is to find methods to ascertain the real cause of accidents which are a result of system failure due to unforeseen interactions among the parts of the system. The objective is to re-engineer the system to prevent accidents in future.

9.7.3 Analysis of the system as a System Theoretic Process Model

System Theoretic Process Analysis (STPA) is done on the control model of the system and factors affecting safety are listed. From an analysis of the electronic interlocking/signalling model given in Fig. 9.2 the following factors affecting safety can be generalized.

1. Control inputs wrong at the computer

2. Requirement not implemented correctly in software

3. Inappropriate or missing control action at the actuator

4. Actuator failure

5. Inadequate or delayed action at the actuator level

6. Component failures or flaws in operation at process level

7. Sensor failure

8. Incorrect or no information at sensor level

9. Some displays missing on the control panel.

From a detailed analysis of these factors, operations likely to cause hazards are listed. The events that may cause hazards are listed below for the electronic interlocking/signalling system on Indian Railways generally termed 'Panel Interlocking'. This is a system where the points and signals are operated individually. Push buttons are provided for selecting each route along with the signal switch. Automatic route setting is not provided. The interlocking is of course electronic, controlled by a computer along with the signalling through outputs. All the operations likely to cause hazards, are constrained by the safety constraints already available in the system evolved over 100 years – signal engineering has evolved in a period of over 100 years and there is no likelihood of accidents as illustrated in the following paragraph. Hence there is no need to introduce additional safety constraints as suggested by Prof. N.G. Leveson.

9.8 Operations or events leading to hazards in the signalling system

The operations or events that can lead to hazards and accidents in the electronic interlocking/ signalling system are listed below and the probable consequences and remedial actions available are mentioned.

a. *A route not intended for reception has been set by the operator by mistake. Probable consequences:* If the unintended line is occupied with wagons or coaches, the route may be set with the concerned points set and locked but the signal for the line (yellow or green) is not lit, as the concerned track relay does not get picked up due to the presence of vehicles on the track circuited portion. The hazard of receiving a train on occupied line is prevented due to the safety constraint already available.

If the rails on this line are being renewed by the permanent way staff, the line will be blocked at either end by physical obstructions, also other precautions like cutting the feed to the signal drive relay, temporarily disconnecting the track relays etc., are taken, so that there is no chance of a hazard.

b. *The requirement of interlocking the level crossing gate in the overlap portion for the home signal has been omitted by mistake in the software, the track circuit in the overlap has of course been included.*

Remedial action: In addition to white box and block box testing of software, functional testing is done by the signal engineers exhaustively at least, at three levels. Verification and validation takes place at least at two levels. This omission is sure to be found out as it is mentioned in the table of control in the last column though insignificantly (LCP – Level Crossing Gate Proving). Correction is done before the system is readied for installation. It can be seen that this sort of omission can be taken care of.

c. *Due to bad contact at the point switch say in the 'RC' portion or a break in the wire connected to the 'reverse' drive relay of the point motor, the feed to the point motor for reversing the point will not be available.*

Result: This results in the point remaining in the normal position even after the point switch is reversed. This does not however result in the loop signal switching to permissive aspect as the normal position is read by the detector circuit. Due to interlocking provided in the software, the loop signal aspect driving relay will not get the feed till the point is reversed and detected as 'reverse'. There is therefore no chance of a hazard taking place i.e. of a signal coming 'off' with the points in the wrong position. The safety constraint is inbuilt.

d. *Due to worn out bushes of the point motor, the motor does not generate the required torque and either the point switch cannot move or if moved may get stuck at the mid position.* With this condition, the detection relays both in normal and reverse positions of the point are de-energized and the same result as detailed in item 'c' above takes place i.e. the point cannot be reversed. The required signal does not take off and there is no chance of a hazard as the constraint is inbuilt in the interlocking software.

e. *Due to the contacts being worn out, the point detection relay does not indicate the correct position of the point.* If the point is in normal position, the normal position detector relay is not energized due to the disconnection in the circuit and the home signal feed for the normal position reception or main line reception is cut off. Hence there is no chance of a hazard due to the inbuilt constraint in the interlocking.

f. If the home signal is switched for receiving the train on main line, stopping at the starter, the 'yellow' aspect is switched on after the route is set and locked. *But due to some defect, the 'yellow' aspect is not indicated on the panel.* The signal at site may be exhibiting 'yellow', but it cannot be confirmed at the operator's panel. The signal switch is now left unoperated and the procedure for receiving the train under 'defective' home signal is activated as provided in the station working rules and the appropriate general rule. The points are to be manually locked and the train piloted by an authorized person with a written memorandum of permission, to the reception area. The manually controlled reception prevents any hazard taking place. But where the traffic density is high, it is advisable to replace the defective indication lamp or if the video monitor is used, to get the defect rectified in quick time. Even if the driver passes the signal by seeing the intermittent 'yellow' aspect, when the operator fiddles with the switch, no hazard is likely to take place, the path being safe.

9.8.1 *Causal analysis based on STAMP (CAST)*

Where an accident has taken place, CAST (Causal Analysis based on STAMP) is to be done wherein the investigation is undertaken to find out the causes by taking the system as a whole and as a control process [4]. This is in contrast to the present methods where the component or subsystem failures are highlighted and the human element is always taken to punish the guilty person as a remedy. This analysis enables the authorities to improve the system by redesigning etc. thereby preventing future accidents. To highlight such methods, an example of the SUBSAFE program adopted by the U.S. Navy is described in the subsequent paragraphs.

9.9 Example of the SUBSAFE program adopted by the U.S. Navy

A brief description of the U.S. Navy's SUBSAFE program is given here to understand the procedure adopted and its relevance to the safety programs being adopted by Railways. For full details the book[4] may be referred to.

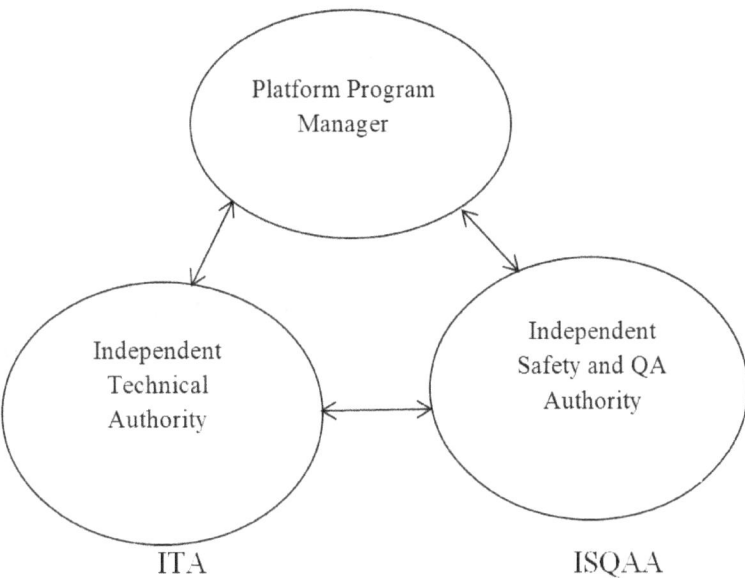

Figure 9.3 Execution of SUBSAFE through three groups

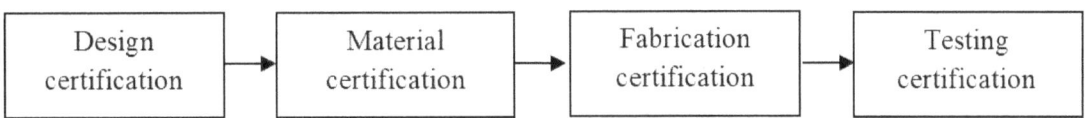

Figure 9.4 Four components of SUBSAFE certification

9.9.1 The SUBSAFE program has been adopted by the U.S. Navy for its nuclear submarines from 1963 onwards after the loss of the nuclear submarine *USS Thresher* at sea on April 10 1963. All the persons on board died. After the program was implemented from 1963 onwards (a few months after the accident), no nuclear submarine was lost at sea for more than fifty years. This is a remarkable achievement.

9.9.2 Reasons for loss of submarine and new remedies based on system life cycle

The main reason for the loss at sea of *USS Thresher* was the presence of deficient silver brazed joints in the salt water piping system which led to flooding in the engine room which could not be controlled, again, due to deficient control gear. The remedial program specific to the submarine was to provide (1) water tight integrity for the submarine's hull and (2) operability and integrity of critical systems to control and recover from a flooding hazard. This ensures that the submarines can surface and return to port safely in an emergency, without being grounded.

To meet the objective of preventing any future accidents, the program covered the stages in the life cycle i.e. design, construction, operation and maintenance. It involved many aspects of submarine development, such as, managerial and technical organization, unique design, material control, fabrication and testing, audit and certification.

The execution of the SUBSAFE program rests anytime on the work of three groups listed in Fig. 9.3. The program manager is responsible for the construction of the submarine within the cost and schedule earmarked, maintaining quality, safety and technical standards set by the other two authorities in the triangle of Fig. 9.3 namely ITA and ISQAA who are responsible for detailed monitoring of the SUBSAFE norms. All the staff comprising these three groups are selected and rigorously trained in the SUBSAFE objectives and constructive engagement between these groups ensures compliance of the norms.

Another vital activity is the certification to be done at four levels of the life cycle which is essential for the upkeep of standards and quality throughout the life of the submarine. The certifications are listed in Fig. 9.4. The certification is based on Objective Quality Evidence (OQE) which is defined as any statement of fact, either qualitative or quantitative, pertaining to the quality of a product or service, based on observations, measurements or tests that can be verified. OQE is taken as a fundamental property to be observed to check whether the requirements have been complied.

Design certification is complete only if the OQE is observed by the competent authority and the technical drawings and other design products are approved by the same authority. *Material certification* involves rigorous inspection of the material and performing chemical and physical OQE including non-destructive testing. *Fabrication certification* is done to verify the industrial processes such as machining, welding and assembly and documentation is done for all the processes with OQE. It is also checked whether the staff doing the high strength steel welding and other skilled operations are qualified and properly trained in the procedures. *Testing certification* is done at the component level, subsystem level and system level throughout the assembly process which includes nondestructive tests and strength testing. For certain components destructive tests are also done on representative samples.

At the end of construction which may take about five years, every submarine obtains its initial SUBSAFE certification. This is done by a competent certification audit team assembled by the Naval Headquarters. The certification is documented and verified throughout the life of the submarine.

9.9.3 Periodical audits

In addition, periodical audits are conducted by a team comprising of staff experienced in SUBSAFE procedures and staff from Naval Headquarters who are concerned with the implementation of SUBSAFE. The audits are both ship specific and functional. The ship specific audit comprises OQE verification of the ship and its suitability to go to sea. The functional audit looks into the policies, procedures and practices followed and whether these comply with the SUBSAFE program requirements.

9.10 Lessons to be learnt in application to electronic interlocking system

While the rigorous procedures adopted in the SUBSAFE program may not be required to be implemented for an electronic interlocking/signalling system, some principles can be adopted to have a more secure and safe system even increasing the reliability of outdoor equipment which is connected to the interlocking system.

9.10.1 The following measures are suggested for improving the robustness and safety aspects of the present electronic interlocking system, though there have been no reports of hazards or accidents due to mal operation of the E.I. system on Indian Railways in the last two decades after installation of such systems which at present exceed 800. The failures reported are mostly of the external gear and peripherals, in which also no unsafe failures have been recorded, due to the inherent fail-safe features in the systems, as pointed out in paragraph 9.8.

(1) Attempts should be made to select the right microprocessor or microcontroller for the E.I. systems being installed on Indian Railways. At present Motorola processors/controllers are being used in most of the systems. As there are a number of microprocessors and microcontrollers in the market, such as, Intel 8086, Motorola 68000, ARM group of Risc microprocessors, Motorola and PIC microcontrollers etc. an appropriate one can be selected. The objective is to reduce the number of operations involved and reduce the amount of Random Access Memory (RAM). Also the operations should be fast as in PowerPC. One or two microprocessors can be selected for the range of routes to be controlled. Factors mentioned in [1] can be considered.

(2) It was reiterated in chapter 4 that a single processor system is more economical than the redundant systems in vogue which are only marginally advantageous. It is recommended that the single processor system is specified for E.I. systems in future, along with a detailed description of the self-diagnostic and self-testing procedures many of which are described in chapter 4.

(3) While all the microprocessor manufacturers are employing some forms of formal methods for verification of the hardware, it is also necessary that the accepting authority (R.D.S.O for IR) should cross verify through various hardware verification methods such as VHDL, MATLAB, Temporal Logic with Model-checking etc. with the help of competent external agencies, so that any hidden fault can be located. These methods should be mentioned in the specification for the system.

(4) As already suggested in Chapter 7, a software engineering group may be formed in RDSO for all the engineering departments, to check the source code of many systems used in operation and maintenance on the Indian Railways. The group can also check the verification and validation of software done by independent agencies before acceptance of new systems, including those imported. More details about the software for safety critical systems and processes of validation will have to be mentioned in the specifications for EI.

(5) Efforts may be made to evolve a specification language in HOL for the railway interlockings and a model verification tool developed for easy verification of the interlockings at the design stage itself. A domain specific language for generating interlocking tables from yard layouts can be considered. Any changes to interlocking in the operation and maintenance stage can also be verified easily for safety etc. under model checking principles.

(6) A central audit team can be formed under the aegis of the RDSO which will check the operation and maintenance of all EI systems on IR by examining the outputs of the data loggers and maintenance terminals from a safety point of view. Also a fraction of the manual logs at such installations can be verified for promptness in attendance to failure messages and the correctness of the remedial action. This is in addition to the scrutiny at the divisional and zonal levels and meets the requirements of a third party supervision. To enable this, the data for each installation should be preserved for at least two years and the audit inspection shall be a minimum of once in a year for each installation.

References:

[1] I.Bate et al "Use of modern processors in safety-critical applications", The Computer Journal – Vol. 44 No. 6 2001.

[2] Bruce Elliott "What is Systems Engineering", presented at University of Birmingham 14 April 2016.

[3] Maurizio Palumbo "Requirements Management for safety critical systems", railwaysignalling. eu UK July 2015.

[4] Nancy G. Leveson, "Engineering a Safer World-systems thinking applied to safety" MIT Press 2012.

[5] J. Santiago and D. Megallon, "critical path method", presented at VDC seminar on 4 Feb. 2009.

A.1 Solid State Interlocking Model

A part of a small scale model of a circular railway corresponding to a station 'X' has been operated by a dedicated microprocessor kit (MTA85–1) incorporating a microprocessor (8085). This included the signals and points at the station which is shown in Fig. A.1. For signals, LEDs (Light Emitting Diodes) were used and three position neutral polar relays were used to simulate point switches. Dual position switches were used to simulate track circuits. A route is set by turning the signal switch either left or right and a push button on the route. The interlocking for which the software is written in assembly language is shown in Table A.1. Approach locking and back locking were also incorporated in this experiment as described in [5](chap.3). The interlocking program running into 7.5 kbytes is stored in EPROM (2716) as shown in the block diagram of the hardware configuration in Fig. A.2. The four EPROM chips are on the expansion board. The switches and push buttons on the control panel, the switches that simulate the track circuit relay contacts and the polarized relay contacts are scanned through five peripheral ICs (8255) and one I/O chip (8155) on the expansion board. The relay coils, signal LEDs and track indication LEDs are also driven through these peripheral chips. The programmable interval timer (8253) in conjunction with I/O chip (8155) and RAM memory were used to generate 120 sec and 60 sec timings for approach locking.

A.2 System Error Detection Proposed

Additional hardware and software were proposed to detect hardware errors in the system and memory. The diagnostic program introduced at the end of the interlocking program will check on stuck-at-0 and stuck-at-1 faults in the system by scanning a square wave with a large period. The memory is checked through generation of '1's and '0's as in barber pole algorithm of H.R. Pinnick. The interlocking requirements or say a line in the interlocking table can be equated to an 'AND' gate or a series of 'AND' gates with sequential inputs. An error in an 'AND' gate can be detected by the presence of synchronous pulses on all the inputs and the output and under certain conditions the presence of a pulse on one of the inputs and the output. This is also confirmed by the concept of multiple boolean difference for an 'AND' gate. In a microprocessor controlled interlocking, the interlocking is in the form of 'firmware' or software but the ability to read a pulse at one of the inputs of a port of a peripheral chip and to generate a pulse at one of the outputs of a port of another peripheral chip is considered a healthy condition of the system, as the inputs in a signalling system are binary and the faults which affect the system are either 'stuck-at-0' or 'stuck-at-1' faults. The absence of a pulse at the output can be considered an erroneous condition of the system. The system should be either totally self-checking i.e. all the faults in the system should be detectable or alternatively partially self-checking and partially fault-secure i.e. the faults which cannot be detected should not affect the output at all so that the output is always predictable.

A.2.1 The hardware configuration for the suggested error-detection method is given in Fig. A.3. The output on a bit of an output port of 8155 of expansion board is connected through on opto-isolator to a bit of input port of peripheral chip 8155. The output from 8155 was programmed to be a square wave of a large period of 1sec and was read by the input port of 8155. Peripheral chips (8255) were used to cater for more inputs in the experiment for duplicated inputs. Any erroneous reading of this output was programmed to feed a high level through a bit of output port of 8155 to the interrupt controller 8259. A hardware detector was also used to check voltage levels and stuck conditions. When this high level interrupt request is received, the interrupt service program is activated and all the red LEDs of signals are switched on. If required, more interrupt requests can be connected to the interrupt controller in the order of priority to cater for various sections of the input/output configuration.

A.2.2 The software for the error detection in assembly language runs at the end of the interlocking program. The '0's and '1's are read every second and in the event of incongruity the interrupt is activated and red LEDs appear.

A.2.3 The random-access memory (RAM) is also checked by means of the barber pole algorithm. A set of barber pole characters (17 in all) are loaded successively in accumulator as well as memory to be checked and compared. If there is a discrepancy, the interrupt is energized.

A.3 Performance of the diagnostic programs

Out of the 116 faults simulated at various points in the system, 73 faults could be detected by means of the hardware detector, interrupt controller and on-board visual display. This comes to about 63%. For the faults not detected, the system was found unaffected. If all the eight bits are connected, the percentage of detection is likely to increase. Regarding checking of RAM, the program performed well in checking stuck_ at faults.

A.3.1 Accident probability evaluation

The fault tree for accident probability for the hardware configuration in Fig. A. 2 is shown in Fig. A.4. The probabilities for the events A to F are as below:

P_A = probability of undetected failure by diagnostic program.

P_B = probability of undetected failure due to mal-operation of detecting hardware/software.

P_C = probability of undetected failure in the period when diagnostic program is not run.

P_D = probability of unwanted route being set during undetected failure.

P_E = probability of train approaching signal during the period

P_F = probability of operator failing to notice abnormality.

The accident probability (per route) as per the fault tree is

$$P_{ACC} = (P_A + P_B + P_C) \times P_D \times P_E \times P_F \quad\text{-- (A.1)}$$

Taking the system failure rate as 16.76×10^{-6} per hour calculated from the failure rates of electronic components, the life of the system as 20 years and various assumptions given in [4] (chap.3) the value of P_{ACC} is calculated to be $P_{ACC} = 2.36 \times 10^{-10}$ per route per day as the factors involved have been taken in a period of 24 h assuming the system is checked every 24 hours for unrevealed failures in the system by means of functional and other tests.

A.4 A method of reducing accident probability

As the accident probability is not within the range of the value recommended by the then U.I.C (International Union of Railways) i.e. 3×10^{-17} (as per question A124) and that by the aviation industry at that time of 3×10^{-18}, a method of reducing the accident probability by means of duplicated input channels and redundant channels is described. Assuming that the system can produce a wrong output only if at least two input paths are faulty or in other words, two inputs are stuck in states not permissible due to the rigorous diagnostic routines, the probability P_U of an unwanted route being set in a system dependent on N inputs including redundant ones can be approximated to

$$P_U \simeq P_S^2 / 2^{N-2} \quad\text{--- (A.2)}$$

where P_S is the probability of any one input being in the wrong state. Analyzing this equation and investigating a method by which a system is not affected by a single fault, it is observed that this is possible only if all the input data channels are duplicated and are fully independent. The inputs can be data from outside functions such as track circuits, points etc. They can also be internal inputs.

A.4.1 The model system was modified as shown in Fig. A.5 to test the above approach. Two different programs have been stored in the EPROM-one scanning the track circuits 1T, 5IT and 01T both through the 8155 and 8255 ICs and the other scanning the same track circuits through an 8255 only. In both programs, when the tracks are free, the '1HG' LED is lit. Otherwise, the '1RG' LED is lit. The LEDs are on a port of another 8255 chip. Various faults including 'stuck-at-0' and 'stuck-at-1' were simulated and the effect on the LEDs was observed for both programs. Either extinguishing the yellow or '1HG' LED or lighting the red or '1RG' LED is considered to be a detectable condition. It was observed that for 124 faults simulated in this system, only 33 were left undetected whereas 54 were left undetected in the original model without duplicated channels.

A.4.2 The probability P_U can be further reduced, if in addition to duplicated input channels, the inputs (N) themselves are increased. For example, for route 1A, routes 1B, 3D and 7G are conflicting and can be proven directly instead of indirectly. Similarly, switch positions 51RWKR and 53RWKR are proved for '1', while 51NWKR and 53NWKR are proved for '0'. It is seen that seven redundant inputs can be included for route 1A. These are in addition to the 16 inputs ordinarily taken. In the model system, route 5E has the lowest number of inputs, two. If these inputs are made redundant and the input channels are duplicated, the probability of this route being set when unwanted using eq. A.2 is

$$P_U = (2 \times 10^{-8})^2/2^{4-2} = 1 \times 10^{-16} \text{ --- (A.3)}$$

Here, it is assumed that $P_S = 2 \times 10^{-8}$ (rounded) the wrong-side failure probability associated with any route in the unmodified solid state portion of the system and is considered to result from any one input being in the wrong state. Once again, assuming that there is a 1 in 100 chance that setting an unwanted route would actually result in an accident, the per route accident probability is $P_U \times 10^{-2} = 1 \times 10^{-18}$. Thus, it can be seen that the accident probability can be reduced to acceptably low levels by suitable manipulation of the inputs and duplication of input channels. In practice, however, full redundancy between inputs and full independence between duplicated channels may not be achievable. This will limit the value of N and the probability P_S may not be fully squared in eq. A.2. Even then, acceptable probability values can be achieved.

A.5 Photographs of the model railway

Photographs of the control panel used and the model circular railway are also shown at Fig. A.6 and Fig. A.7.

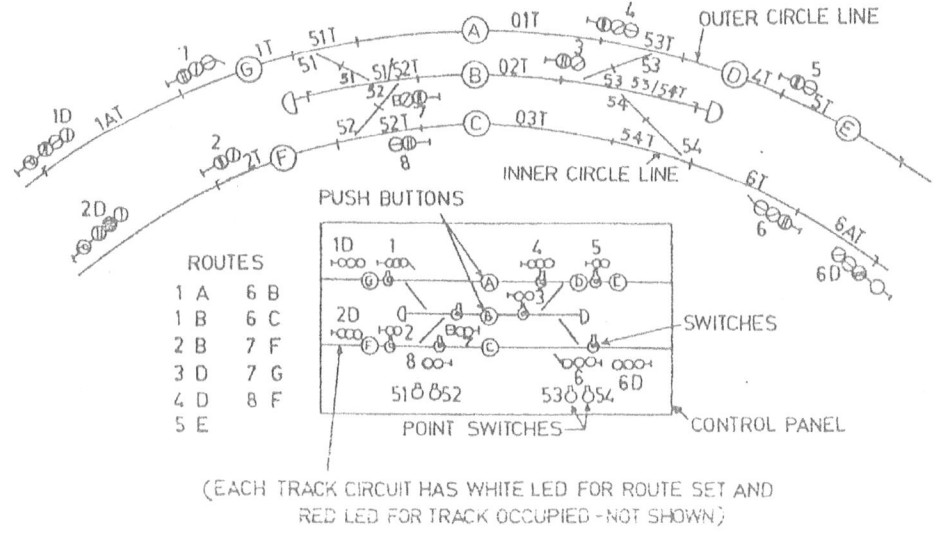

SIGNALLING PLAN & CONTROL PANEL FOR STATION 'X'

FIG.A.1

Figure A.1 Signalling plan and control panel diagram for model station 'X'

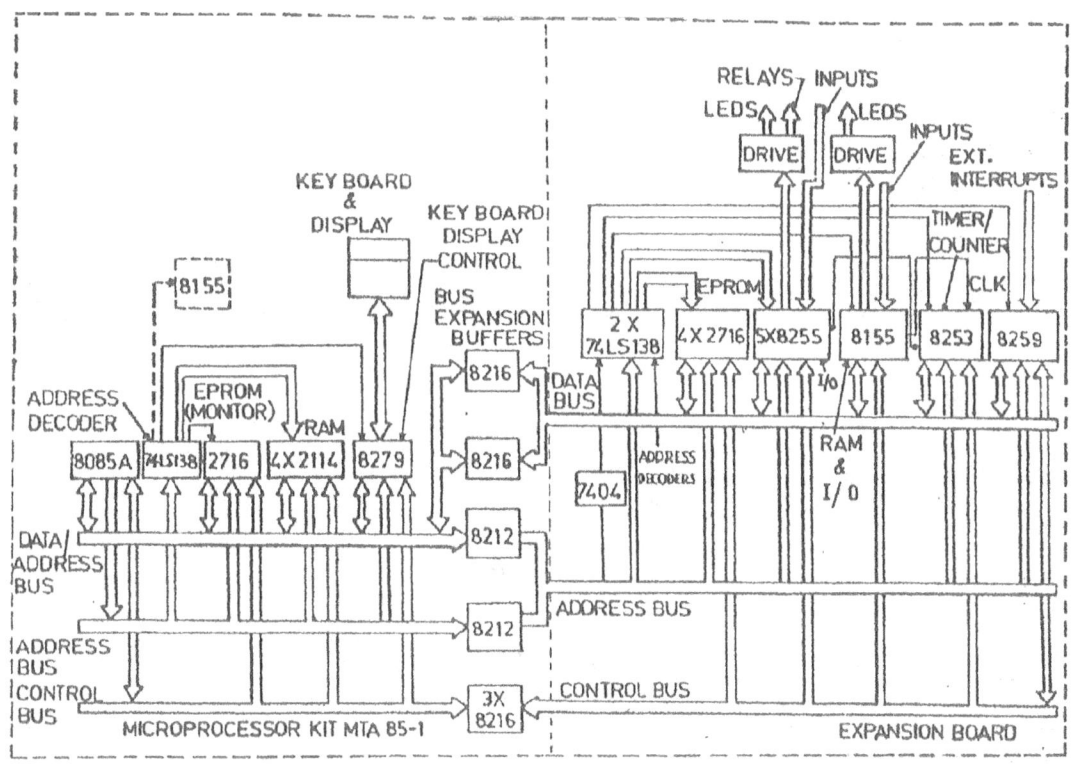

Figure A.2 Hardware configuration of microprocessor based control system

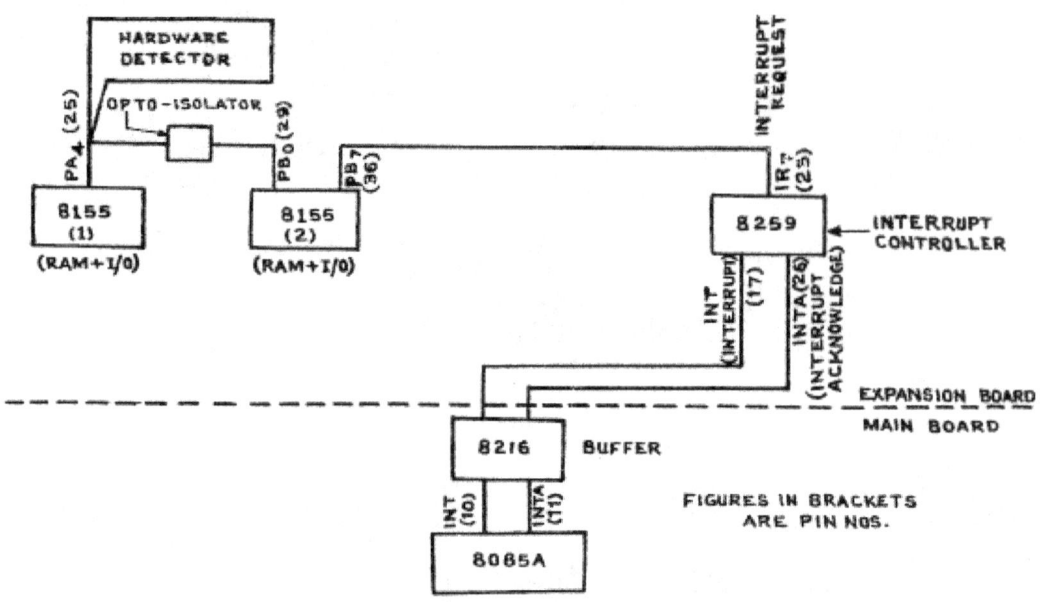

Figure A.3 Hardware configuration for the proposed error detection method

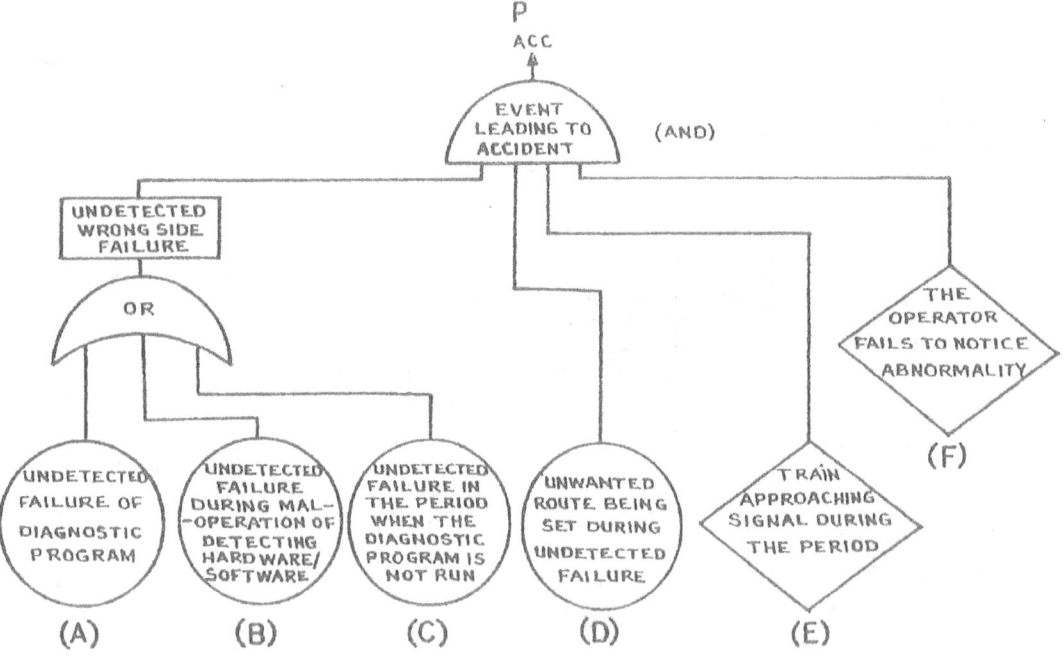

Figure A.4 Fault tree for accident probability

Table A.1. Control table for station 'X'

S. No.	Routes	Description	Locks and Detects Points	Locks Routes	Controlled by Tracks	Approach Tracks	Back Lock Tracks	Remarks
1.	1A	Reception to line 1 from outer line	51N, 53N	1B, 3D, 7G	1T, 51T, 01T, 53T	1AT, 120 SEC	1T, 51T	1DG with 4DG & 5DG
2.	1B	Reception to line 2 from outer line	51R, 52N, 53R, 54N	1A, 2B, 4D, 6B, 7E, 7G	1T, 51T, 51/52T 02T, 53/54T, 53T	1AT, 120 SEC	1T, 51T 51/52T	-
3.	2B	Reception to line 2 from inner line set to sand hump	51N, 52R, 53N, 54N	1B, 3D, 6B, 6C, 7E, 7G, 8F	2T, 52T, 51/52T, 02T, 53/54T	120 SEC	2T, 52T 51/52T	-
4.	3D	Despatch from line 2 to outer line	53R, 54N	1A, 2B, 4D, 6B 7E, 7G	53/54T, 53T, 4T	02T, 60 SEC	53/54T, 53T	-
5.	4D	Despatch from line 1 to outer line	53N	1B, 3D	53T, 4T	01T, 1NW(1Aset) 120 SEC	53T	4DG with 5DG
6.	5E	Advanced dispatch to outer line	-	-	5T	-	-	-
7.	6B	Reception to line 2 from inner line	51N, 52N, 53N, 54R	1B, 2B, 3D, 6C 7E, 7G	6T, 54T, 53/54T 02T, 51/52T	6AT, 120 SEC	6T, 54T 53/54T	SET TO SANDHU-MP
8.	6C	Reception to line 3 from inner line	52N, 54N	2B, 6B, 7F	6T, 54T, 03T, 52T	6AT, 120 SEC	6T, 54T	6DG WITH 8 DG
9.	7F	Despatch to inner line from line 2	51N, 52R	1B, 2B, 3D, 6B, 6C, 7G, 8F	51/52T, 52T, 2T	02T, 60 SEC	51/52T, 52T	-
10.	7G	Despatch to outer line from line 2	51R, 52N	1A, 1B, 2B, 3D, 6B, 7F	51/52T, 51T, 1T	02T, 60 SEC	51/52T, 51T	-
11.	8F	Depatch to inner line from line 3	52N	2B, 7F	52T, 2T	03T, 120 SEC	52T	-

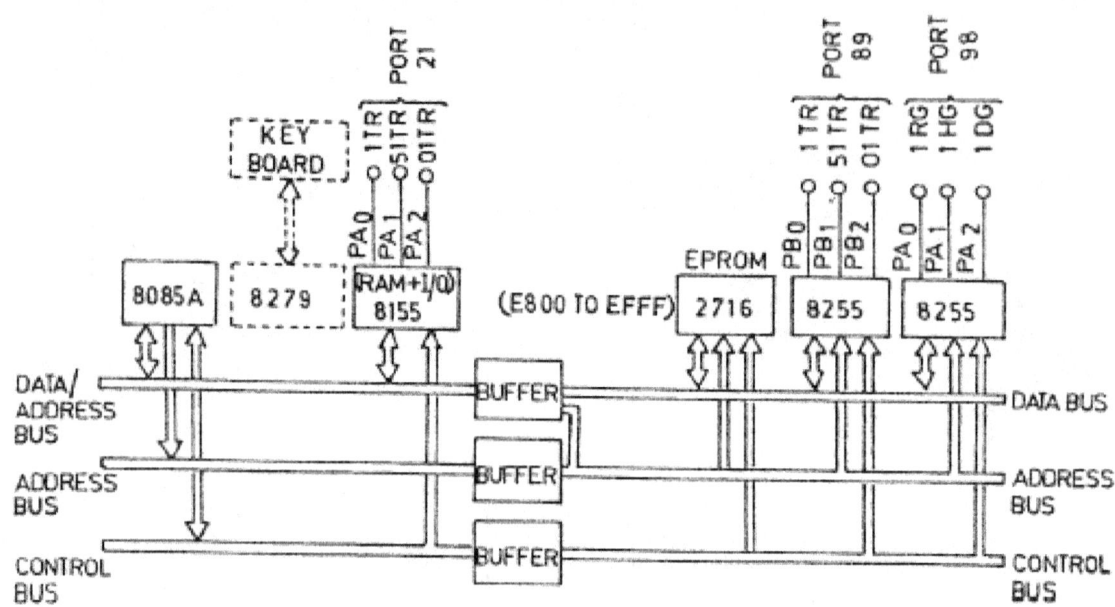

Figure A.5 Experimental system with duplicate input channels

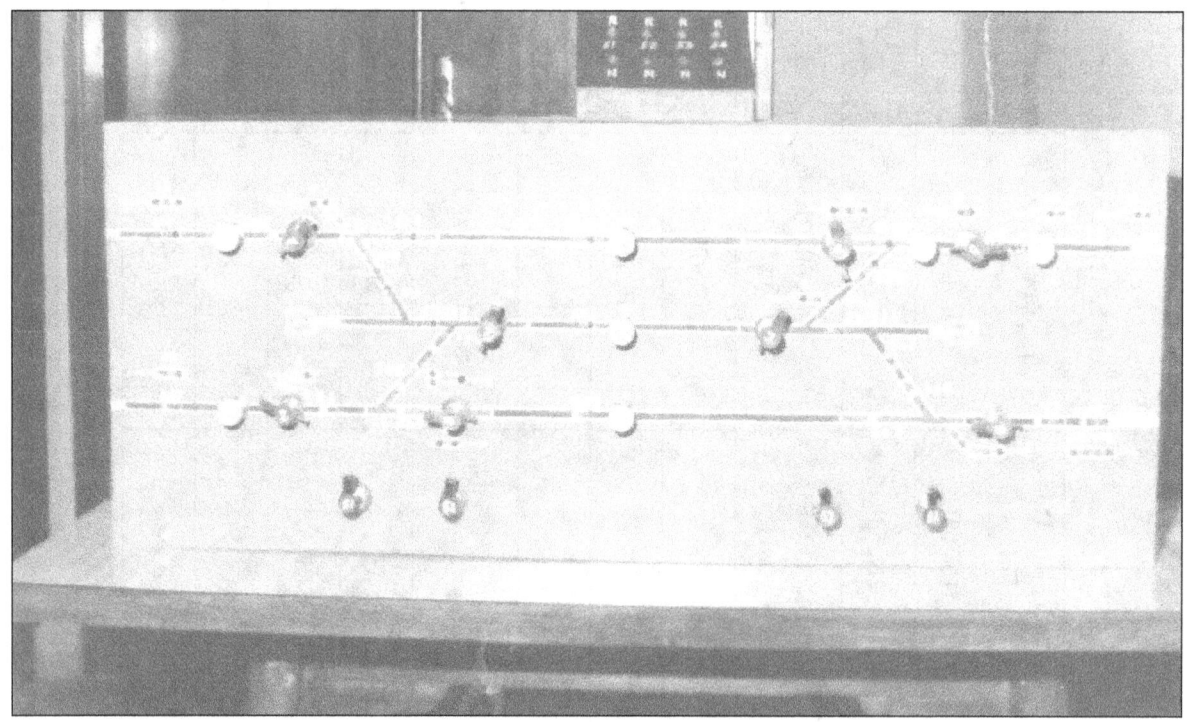

Figure A.6 Control panel of Station 'X'(photo)

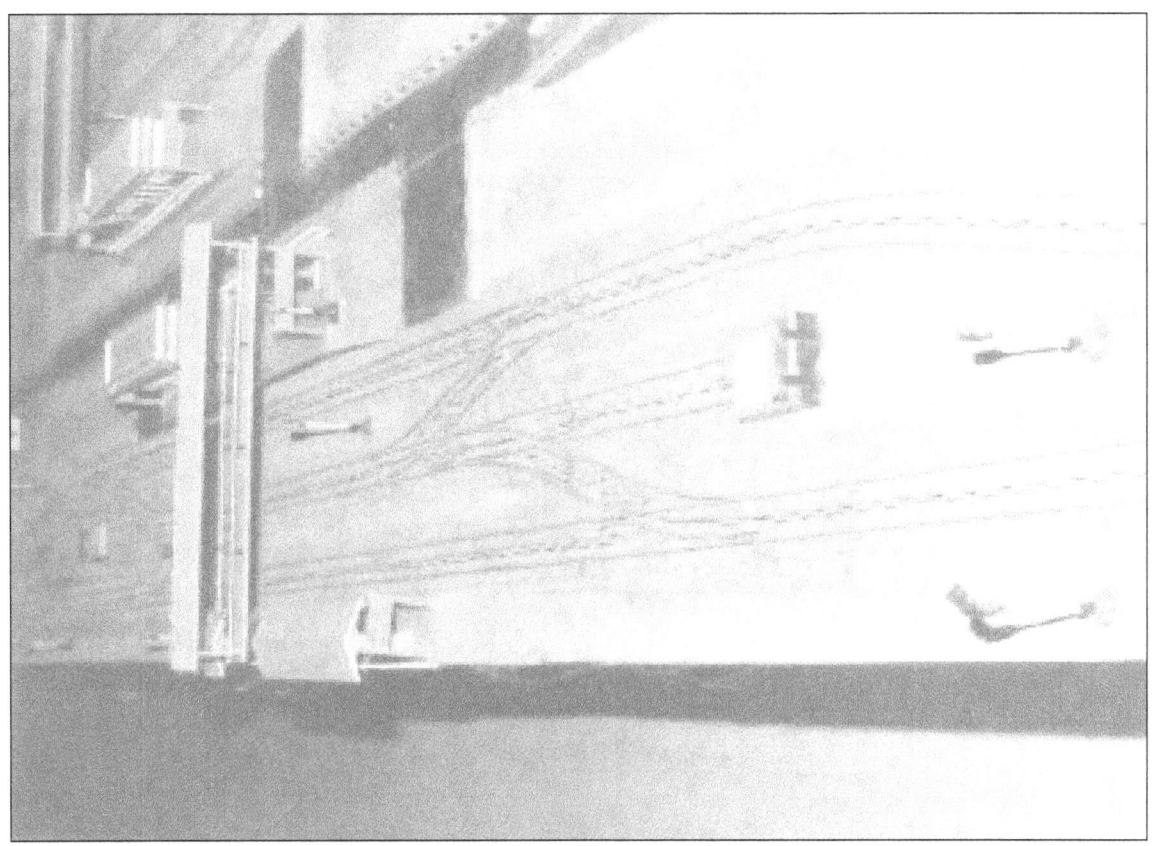

Figure A.7 Model of station 'X' – A view

ELECTRONIC INTERLOCKINGS INSTALLED EARLIER ON INDIAN RAILWAYS

B.1 ESA11–1R of AZD Praha

The ESA11-IR E.I. was supplied by M/s.AZD Praha of Czech Republic. It has a 2 out of 2 architecture as shown in Fig. B.1. It provides for hot stand by at the vital processor level. The processors operate with two independent softwares to eliminate common mode failures as claimed by the supplier. The comparator monitors the instantaneous outputs and in case of non-agreement the vital cutoff relay drops and the power supply to vital output drivers is cut off, leading to a safe state. Any failure in the working pair of CPUs will result in smooth changeover to hot stand by pair of CPUs. The Object Controllers also have a 2 out of 2 architecture with independent softwares. Power cutoff feature is also inbuilt for the output modules. The system can cater for both VDU panel and conventional control panel but VDU panel is recommended by the firm due to its inherent advantages of providing additional information to the operating personnel. There is provision for connecting Maintenance Terminal and Event Data Logger for monitoring of events and faults. Remote control or Central Traffic Control (CTC) operations are provided for through the serial interface.

B.1.1 In case of yard alterations when required, field modules can be added or removed as per the need and software for the remodelled yard can be programmed by the use of graphic user interface and menu driven software interlocking. 11 nos. EI systems of this type were installed on one zonal railway.

B.2 VHLC-EI of M/s GE

The block diagram of VHLC-EI of M/s GE is drawn in Fig. B.2. This is a 2 out of 2 check redundant system. The VHLC (Vital Harmon Logic Controller), with the chassis portion shown in dotted lines, comprises the Vital Logic Processor (VLP) which is a dual processor system with a comparator checking the congruence. Its vital application software is stored in EPROMS of the Site Specific Module (SSM) based on vital logic equations appropriate to the controlled yard. The application software is compiled by the Application Compiler Editor (ACE) in a Windows based computer with the input logic equations fed by the site engineer. The Auxiliary Communications Processor (ACP) processes non-vital logic and handles serial data communication. It communicates with additional VHLCs through RS232 interface and with the Operator and/or Maintenance PC through RS232 or RS485 interfaces.

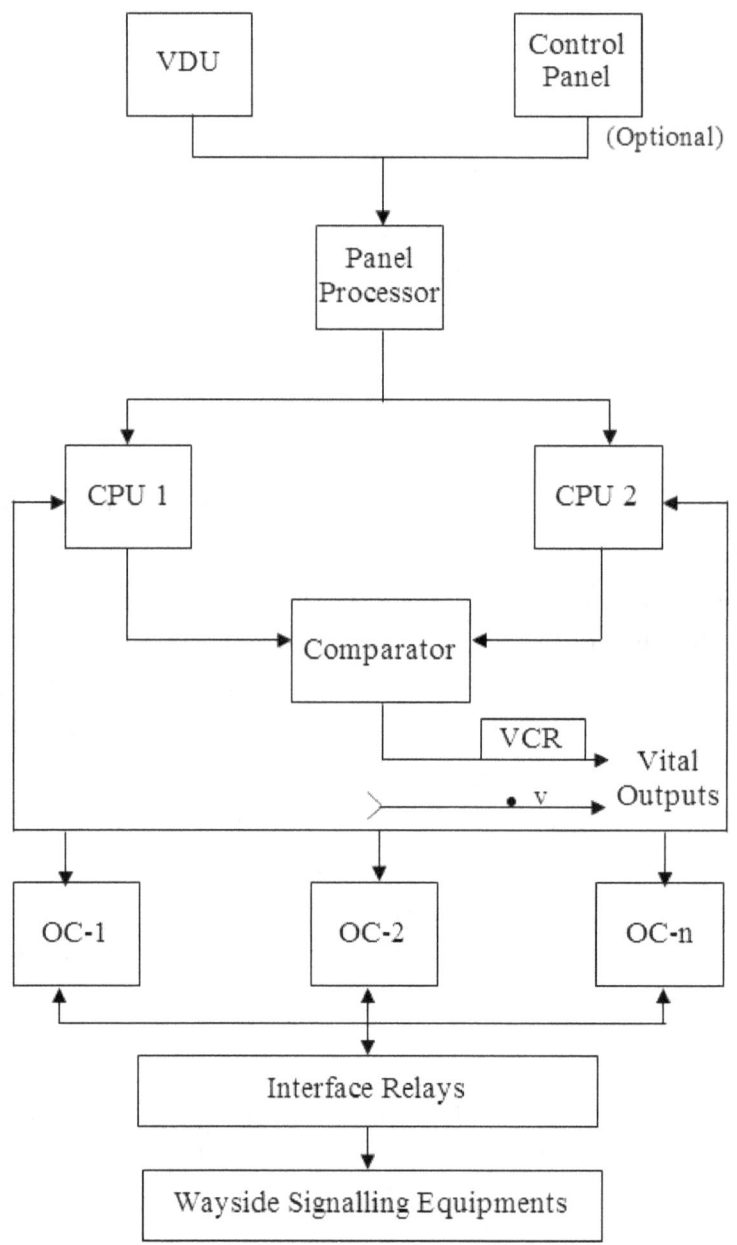

VCR: Vital Cut off Relay   OC: Object Controller

VDU: Visual Display Unit   CPU: Central Processing Unit

Figure B.1 Architecture of ESA11-IR E.I. SYSTEM (AZD PRAHA)

It is connected to the local control panel through a current loop adaptor which is a 2/4 wire converter through switches. Vital inputs and outputs are buffered to the Vital Logic Processor through the Input/Output modules and the vital output relays can be driven at the specified voltages. 30 nos. EI systems of this type were installed on various Zonal Railways.

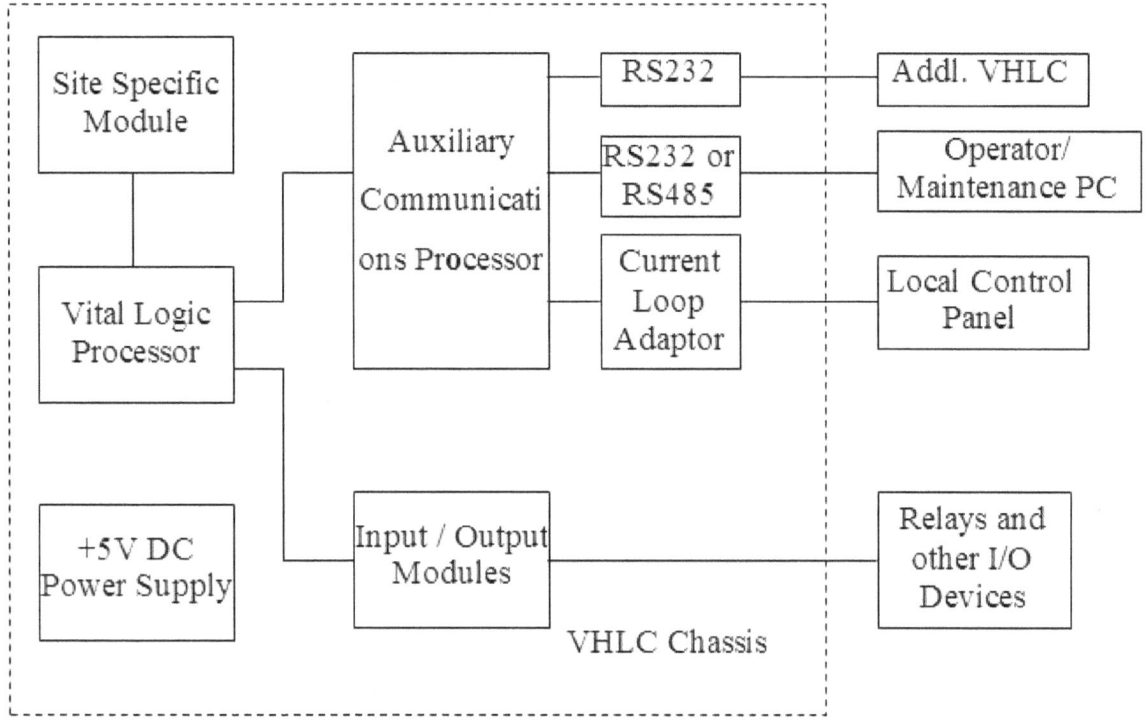

Figure B.2 Block Diagram of VHLC-Electr. Interlocking (M/s GE)

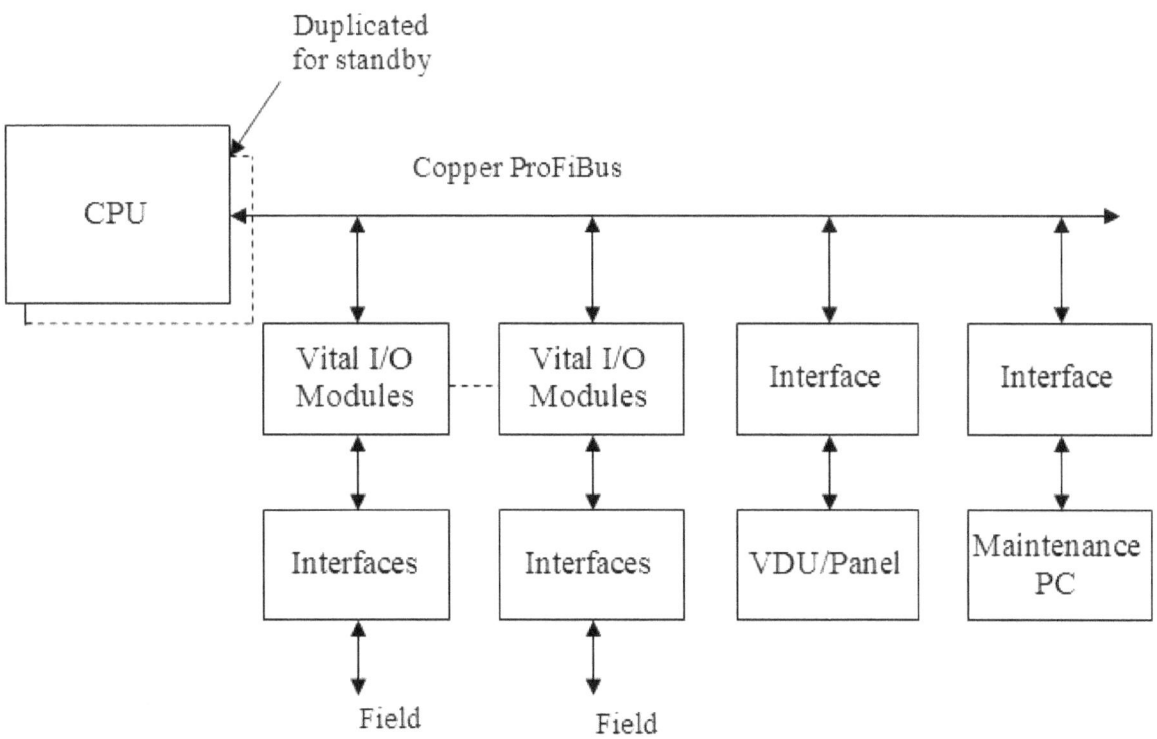

ProFiBus : Process Field Bus

Figure B.3 Block Diagram of SIMIS S E.I. System

B.3 SIMIS S EI System

SIMIS S is a single processor Electronic Interlocking system manufactured by Siemens Ltd. of Germany. SIMIS is a brand name for the system and is an acronym of SIcheres MIkrocomputer System (German for safe microcomputer system). This system works with a hot standby so that there is negligible interruption in case of failures. Operation of the system can be through a conventional control panel of VDU (Visual Display Unit) console. The block diagram of the system is given in Fig. B.3. Copper 'ProFiBus' is used for connecting the various modules if the distance is within 200 m and if it exceeds, optic fibre cable with the required interfaces is generally used. The serial links can be through the communication processor, RS 232/422, TCP/IP, Ethernet and the optic fibre interfaces wherever required.

B.3.1 The executing program is stored in a flash EPROM which can be plugged in to the memory module of the CPU. Also both CPUs have back-up batteries that provide a backup of the program in RAM. The interlocking logic for the yard is incorporated in the executing program by compiling of the STEP 7 programming language which is a proprietary item of this firm. This system does not have the vital cut off feature but indicates 'RED' when hardware/software faults occur and STOP (in yellow) when the CPU does not process the user program. However there is a provision for attaining a safe state in 1.5 secs if a hazardous event occurs. 23 nos. installations of this type are now existing on the main line railways.

/* PROGRAM FOR SIMULATION OF STATION "Y" */

```
#include<stdio.h>
#include<conio.h>
#include<graphics.h>
#include<stdlib.h>
#include<dos.h>
#include<string.h>
#include<iostream.h>

int points51=0; /* normal position of points, */
int points53=0; /* point locks and track circuits */
int pointlock51=0;
int pointlock53=0;
int trackcct1T=0;
int trackcct1AT=0;
int trackcct51T=0;
int trackcct01T=0;
int trackcct01AT=0;
int trackcct53T=0;
int trackcct2T=0;
int trackcct4T=0;
int LCP=0; /* level crossing gate plunger */
int dmcheck=0; /* parameter to check points, track circuits & LCP on down main line */
int DRP=0; /* down reception plunger */
int BL=0; /* block instrument towards station 'Z'- indication */
int DG4=0; /* down advanced starter green indication */
int DG2=0; /* down main starter green indication */
char route1[5]={"*1A*"}; /* characters to indicate inputs */
char route2[5]={"*1B*"}; /* for the ten functions */
char route3[5]={"*2E*"};
char route4[5]={"*3E*"};
char route5[5]={"*4F*"};
```

```c
char routechk1[5]={"1A*B"};
char routechk2[5]={"1B*A"};
char route1Abklock[5]={"1ABL"};
char route1Aapplock[5]={"1AAL"};
char route1Amlthrecep[5]={"MLTH"};

void checkdnmain()
  {
   if(point51==0 && point53==0 && trackcct1T==0 && trackcct1AT==0 && trackcct51T==0 && trackcct01T==0
     && trackcct01AT==0 && trackcct53T==0 && LCP==0)
    {
     dmcheck=1;
    }
     else
    {
     dmcheck=0;
    }
  }
void setroute1A();/* functions for setting the five */
void setroute1B();/* routes available */
void setroute2E();
void setroute3E();
void setroute4F();
void abnorm1();/* after setting route 1A try setting route 1B */
void abnorm2();/* after setting route 1B try setting route 1A */
void dnmainbklock();/demonstrate backlocking for route 1A */
void dnmainapplock(); /* demonstrate approach locking for route 1A */
void mlthrecep();/* demonstrate signaling for a through train */

main ()
{
int gd=9, gm=VGAHI;
initgraph(&gd,&gm, "C:\\TURBOC3\\bgi");
setbkcolor(BLACK);
outtextxy(200, 380,"SIGNALLING PLAN FOR STATION Y");
outtextxy(240, 390,"AND TABLE OF CONTROL");
outtextxy(10, 90,"<- STATION X");
outtextxy(550, 100,"STATION Z ->");
outtextxy(178, 123,"(upside not fully illustrated)");
rectangle(310, 30, 350, 50); /* control panel housing */
circle(330, 42, 2);/* controller */
circle(340, 45, 1); /* DRP-down loop reception plunger */
circle(345, 45, 1); /* LCP-level crossing gate plunger */
outtextxy(280, 20,"CONTROL PANEL");
outtextxy(352, 42,"<- DRP, LCP");
setlinestyle(0, 0, 3);
```

```
line(320, 35, 340, 35);/* control panel */
setlinestyle(0, 0, 1);
line(0, 70, 640, 70); /* down main line */
line(200, 90, 450, 90); /* down loop line */
line(190, 73, 203, 87); /* 1st X'over */
line(447, 87, 460, 73); /* 2nd X'over */
line(0, 120, 640, 120); /* up main line */
line(170, 140, 420, 140); /* up loop line */
line(160, 123, 173, 137); /* 2nd X'over */
line(417, 137, 430, 123); /* 1st X'over */
setlinestyle(1, 0, 1);
line(460, 118, 490, 72); /* emergency X'over */
outtextxy(480, 90,"<- em'ncy"); /* between up and down lines */
outtextxy(496, 97,"X over");
sector(10, 70, 80, 100, 8, 3); /* pointer to down warning board */
sector(50, 70, 80, 100, 8, 3); /*pointer to down distant signal */
outtextxy(60, 73,"-->");
outtextxy(85, 73,"DOWN");
sector(130, 70, 80, 100, 8, 3);/* pointer to down home signal */
sector(440, 70, 80, 100, 8, 3);/* pointer to down main starter */
sector(440, 90, 80, 100, 8, 3);/* pointer to down loop starter */
sector(560, 70, 80, 100, 8, 3);/* pointer to down advanced starter */
sector(610, 120, 260, 280, 8, 3);/* pointer to up warning board */
sector(560, 120, 260, 280, 8, 3);/* pointer to up distant signal */
sector(480, 120, 260, 280, 8, 3);/* pointer to up home signal */
sector(180, 120, 80, 100, 8, 3);/* pointer to up main starter */
sector(180, 140, 260, 280, 8, 3);/* pointer to up loop starter */
sector(60, 120, 260, 280, 8, 3);/* pointer to up advanced starter */
setlinestyle(0, 0, 1);
line(10, 67, 10, 59);
line(10, 59, 20, 59);
rectangle(20, 56, 36, 62);/* down warning board */
line(610, 123, 610, 131);
line(610, 131, 600, 131);
rectangle(584, 128, 600, 134);/* up warning board */
line(50, 67, 50, 59);
line(50, 59, 60, 59);
circle(63, 59, 3);/*'P' indication */
line(66, 59, 71, 59);
circle(74, 59, 3);/* down distant signal */
setfillstyle(SOLID_FILL, YELLOW);
floodfill(74, 59, WHITE);
line(560, 123, 560, 131);
line(560, 131, 550, 131);
circle(547, 131, 3);/* 'P' indication */
line(544, 131, 539, 131);
```

```
circle(536, 131, 3);/* up distant signal */
setfillstyle(SOLID_FILL, YELLOW);
floodfill(536, 131, WHITE);
circle(28, 59, 3);/* on down warning board */
line(22, 56, 22, 62);
line(34, 56, 34, 62);
circle(592, 131, 3);/* on up warning board */
line(586, 128, 586, 134);
line(598, 128, 598, 134);
circle(80, 59, 3);/* down distant green */
circle(86, 59, 3);/* down distant double yellow */
circle(530, 131, 3);/* up distant green */
circle(524, 131, 3);/* up distant double yellow */
outtextxy(61, 49,"P");
outtextxy(547, 136,"P");
line(130, 67, 130, 59);
line(130, 59, 151, 59);
circle(154, 59, 3);/* down home signal */
setfillstyle(SOLID_FILL, RED);
floodfill(154, 59, WHITE);
line(480, 123, 480, 131);
line(480, 131, 459, 131);
circle(456, 131, 3);/* up home signal */
setfillstyle(SOLID_FILL, RED);
floodfill(456, 131, WHITE);
circle(160, 59, 3);/* down home yellow */
circle(166, 59, 3);/* down home green */
circle(450, 131, 3);/* up home yellow */
circle(444, 131, 3);/* up home green */
rectangle(172, 59, 169, 65);/* down home route indicator */
rectangle(441, 131, 438, 137);/* up home route indicator */
line(440, 67, 440, 59);
line(440, 59, 461, 59);
circle(464, 59, 3);/* three lamps of */
circle(470, 59, 3);/* down main line starter */
circle(476, 59, 3);
line(440, 94, 440, 98);
line(440, 96, 458, 96);
circle(462, 96, 3);/* two lamps of */
circle(468, 96, 3);/* down loop starter */
setlinestyle(1, 0, 1);/* dotted lines to down loop */
line(440, 87, 440, 83);/* starter location */
line(440, 83, 432, 83);
line(432, 83, 432, 96);
line(432, 96, 440, 96);
setlinestyle(0, 0, 1);
```

```
line(560, 67, 560, 59);
line(560, 59, 581, 59);
circle(584, 59, 3);/* two lamps of */
circle(590, 59, 3);/* down advanced starter */
circle(572, 53, 2);/* indication of block instrument */
outtextxy(565, 42,"BL");
setfillstyle(SOLID_FILL, RED);
floodfill(464, 59, WHITE);/* normal red lights */
floodfill(462, 96, WHITE);/* of down starters */
floodfill(584, 59, WHITE);
line(180, 117, 180, 109);
line(180, 109, 159, 109);
circle(156, 109, 3);/* three lamps of */
circle(150, 109, 3);/* up main line starter */
circle(144, 109, 3);
line(180, 143, 180, 151);
line(180, 151, 161, 151);
circle(158, 151, 3);/* two lamps of */
circle(152, 151, 3);/* up loop line starter */
line(60, 123, 60, 131);
line(60, 131, 39, 131);
circle(36, 131, 3);/* two lamps of */
circle(30, 131, 3);/* up advanced starter */
setfillstyle(SOLID_FILL, RED);
floodfill(156, 109, WHITE);/* normal red lights */
floodfill(158, 151, WHITE);/* of up starters */
floodfill(36, 131, WHITE);
outtextxy(74, 46,"1D");/* all signal numbers up and */
outtextxy(517, 137,"10D");/* down side */
outtextxy(157, 46,"1");
circle(157, 38, 6);/* switch for signal 1 */
rectangle(155, 29, 159, 32);
outtextxy(444, 138,"10");
outtextxy(467, 46,"2");
circle(467, 38, 6);/* switch for signal 2 */
rectangle(465, 29, 469, 32);
outtextxy(459, 101,"3");
circle(447, 108, 6); /* switch for signal 3 */
rectangle(445, 99, 449, 102);
outtextxy(597, 54,"4");
circle(597, 38, 6);/* switch for signal 4 */
rectangle(595, 29, 599, 32);
outtextxy(133, 106,"9");
outtextxy(141, 148,"8");
outtextxy(19, 128,"7");
outtextxy(500, 112,"UP<--");
```

```
outtextxy(138, 67,"L");/* track circuit borders */
line(183, 67, 183, 73);/* near down home signal-end 1AT */
line(167, 67, 167, 73)/* end 1T */
line(209, 67, 209, 73);/* end 51T */
line(209, 87, 209, 93);/* track circuit borders(end 51T on loop line) */
line(206, 87, 209, 87);/* on down lines */
outtextxy(312, 73,"dn");
line(325, 67, 325, 73);/* end 01T */
outtextxy(341, 73,"main");
outtextxy(266, 93,"dn loop");
line(444, 67, 444, 73);/* end 01AT */
line(444, 86, 444, 93);/* 01AT on loop line */
line(444, 86, 445, 86);
line(464, 67, 464, 73);/* end 53T */
circle(333, 70, 5);/* knob "A" on down main line */
outtextxy(330, 56,"A");
circle(333, 90, 5);
outtextxy(330, 97,"B");/* knob "B" on down loop line */
outtextxy(150, 73,"1T");/* track circuit numbers */
outtextxy(168, 79,"1AT");/* down side */
outtextxy(201, 75,"51T");
outtextxy(195, 94,"51");/* 1st 'X'over number */
outtextxy(262, 73,"01T");/* down main line */
outtextxy(382, 73,"01AT");/* track circuit numbers */
outtextxy(454, 85,"53");/* 2nd 'X'over number */
outtextxy(432, 73,"53T");
outtextxy(515, 73,"2T");
circle(500, 70, 5);/* knob 'E' for departure */
outtextxy(498, 77,"E");/* towards station "Z" */
circle(600, 70, 5);/* knob 'F' for entry into */
outtextxy(598, 77,"F");/* block section */
line(565, 67, 565, 73);/* limits of track circuit */
line(579, 67, 579, 73);/* FVT "4T" */
line(576, 67, 579, 67);
outtextxy(567, 76,"4T");
setlinestyle(1, 0, 1);/* indication of level crossing */
line(449, 65, 449, 45);/* gates on both sides of the */
line(459, 65, 459, 45);/* station yard, the gates are */
line(465, 125, 465, 145);/* interlocked through an */
line(475, 125, 475, 145);/* electric lever lock */
circle(449, 65, 2);/* operated from the central */
circle(465, 125, 2);/* control panel */
setlinestyle(0, 0, 1);
line(449, 63, 460, 65);
line(449, 67, 460, 65);
line(465, 123, 476, 125);
```

```
line(465, 127, 476, 125);
outtextxy(460, 148,"L C");/* pointer to LC gate */
outtextxy(455, 157,"GATE");
outtextxy(200, 170,"TABLE OF CONTROL(DOWN SIDE)");
line(50, 180, 560, 180);/* table constructed with */
line(50, 180, 50, 320);/* the following lines  */
line(560, 180, 560, 320);/* as borders */
line(50, 320, 560, 320);
line(50, 220, 560, 220);
line(130, 180, 130, 320);
line(200, 180, 200, 320);
line(240, 180, 240, 320);
line(320, 180, 320, 320);
line(320, 200, 490, 200);
line(405, 200, 405, 320);
line(490, 180, 490, 320);
outtextxy(73, 196,"ROUTE");
outtextxy(140, 185,"LOCKS &");
outtextxy(140, 195,"DETECTS");
outtextxy(143, 203,"POINTS");
outtextxy(201, 190,"l'cks");
outtextxy(201, 205,"r'tes");
outtextxy(245, 190,"CONTR'LED");
outtextxy(248, 205,"BY TRACKS");
outtextxy(350, 188,"ROUTE HELD BY");
outtextxy(322, 208,"app tracks");
outtextxy(410, 208,"blk tracks");
outtextxy(500, 195,"REMARKS");
outtextxy(85, 225,"1A");
outtextxy(137, 225,"51N, 53N");
outtextxy(205, 225,"----");
outtextxy(242, 225,"1T, 1AT, 51T");
outtextxy(245, 235,"01T, 01AT");
outtextxy(260, 245,"53T");
outtextxy(333, 225,"120 sec");
outtextxy(410, 225,"1T, 1AT, 51T");
outtextxy(493, 225,"2DGforDG");
outtextxy(500, 235,"LCP");
outtextxy(85, 255,"1B");
outtextxy(137, 255,"51R, 53R");
outtextxy(205, 255,"----");
outtextxy(242, 255,"1T, 1AT, 51T");
outtextxy(260, 265,"53T");
outtextxy(333, 255,"120 sec");
outtextxy(410, 255,"1T, 1AT, 51T");
outtextxy(498, 255,"LCP, DRP");
```

```
outtextxy(85, 275,"2E");
outtextxy(155, 275,"53N");
outtextxy(205, 275,"----");
outtextxy(255, 275,"53T, 2T");
outtextxy(330, 275,"01T, 01AT");
outtextxy(333, 285,"120 sec");
outtextxy(440, 275,"53T");
outtextxy(493, 275,"4DGforDG");
outtextxy(500, 285,"LCP");
outtextxy(85, 295,"3E");
outtextxy(155, 295,"53R");
outtextxy(205, 295,"----");
outtextxy(255, 295,"53T, 2T");
outtextxy(338, 295,"60 sec");
outtextxy(440, 295,"53T");
outtextxy(515, 295,"LCP");
outtextxy(85, 305,"4F");
outtextxy(150, 305,"----");
outtextxy(205, 305,"----");
outtextxy(270, 305,"4T");
outtextxy(350, 305,"----");
outtextxy(435, 305,"----");
outtextxy(493, 305,"blk inst");
char route[5];
int x1;
int x2;
int x3;
int x4;
int x5;
int x6;
int x7;
int x8;
int x9;
int x10;
delay (4000);
gets(route);
x1=strcmp(route1, route);
x2=strcmp(route2, route);
x3=strcmp(route3, route);
x4=strcmp(route4, route);
x5=strcmp(route5, route);
x6=strcmp(routechk1, route);
x7=strcmp(routechk2, route);
x8=strcmp(route1Abklock, route);
x9=strcmp(route1Aapplock, route);
x10=strcmp(route1Amlthrecep, route);
```

```
switch(x1)
case 0:
        if(x1==0)
        {
            setroute1A();
        }
switch(x2)
 case 0:
        if(x2==0)
        {
            setroute1B();
        }
switch(x3)
 case 0:
        if(x3==0)
        {
            setroute2E();
        }
switch(x4)
 case 0:
        if(x4==0)
        {
            setroute3E();
        }
switch(x5)
 case 0:
        if(x5==0)
        {
            setroute4F();
        }
switch(x6)
 case 0:
        if(x6==0)
        {
            abnorm1();
        }
switch(x7)
 case 0:
        if(x7==0)
        {
            abnorm2();
        }
switch(x8)
 case 0:
        if(x8==0)
        {
```

```
                        dnmainbklock();
                }
    switch(x9)
     case 0:
                if(x9==0)
                {
                    dnmainapplock();
                }
    switch(x10)
     case 0:
                if(x10==0)
                {
                    mlthrecep();
                }
 getch();
 closegraph();
 return0;
}
void setroute1A()
{
 rectangle(163, 36, 167, 39);
 setfillstyle(SOLID_FILL, CYAN);/* switch 1 turned right */
 floodfill(157, 38, WHITE);/* and knob A pressed */
 floodfill(164, 37, WHITE);/* shown in color cyan */
 floodfill(330, 68, WHITE);
 floodfill(335, 71, WHITE);
 checkdnmain();
 delay(4000);
 if(dmcheck==1)
 {
  pointlock51=1;
  pointlock53=1;
  setfillstyle(SOLID_FILL, BLACK);/* blanking red of down home */
  floodfill(154, 59, WHITE);/* signal and switching on */
  setfillstyle(SOLID_FILL, YELLOW);/* yellow with route set */
  floodfill(160, 59, WHITE);
  setfillstyle(SOLID_FILL, BLACK);/* yellow of down distant */
  floodfill(74, 59, WHITE);/* blanked and green lit with */
  setfillstyle(SOLID_FILL, GREEN);/* points 51N & signal 1 yellow */
  floodfill(80, 59, WHITE);
  rectangle(138, 68, 440, 70);/* track circuited portion of down home */
  setfillstyle(SOLID_FILL, YELLOW);
  floodfill(160, 69, WHITE);/* corresponding to 1T */
  floodfill(175, 69, WHITE);/* corresponding to 1AT */
  floodfill(200, 69, WHITE);/* corresponding to 51T */
  floodfill(320, 69, WHITE);/* corresponding to 01T */
```

```
  floodfill(327, 69, WHITE);/* corresponding to 01AT */
  floodfill(400, 69, WHITE);/* excluding knob A */
if(DG2==1)
 {
  setfillstyle(SOLID_FILL, BLACK);
  floodfill(160, 59, WHITE);/* yellow blanked and */
  setfillstyle(SOLID_FILL, GREEN);/* down home turning green */
  floodfill(166, 59, WHITE);/* when down main starter turns green */
 }
 }
  else
  {
    cout<<"\n conditions not present for route 1A \n";
  }
};
void setroute1B()
{
  rectangle(163, 36, 167, 39);
  setfillstyle(SOLID_FILL, CYAN);/* switch 1 turned right */
  floodfill(157, 38, WHITE);/* and knob B pressed */
  floodfill(164, 37, WHITE);/* shown in cyan color */
  floodfill(330, 88, WHITE);
  floodfill(335, 92, WHITE);
  if(pointlock51==0 && pointlock53==0 && trackcct1T==0 && trackcct1AT==0    && trackcct51T==0 && DRP==0 &&
trackcct53T==0 && LCP==0 )
  {
    point51=1;
    point53=1;
    pointlock51=1;
    pointlock53=1;
    arc(175, 80, 25, 40, 17);
    arc(205, 60, 205, 220, 17);/* reversing of points */
    arc(215, 80, 205, 220, 15);/* 51 & 53 for down loop home */
    arc(435, 70, 285, 306, 22);
    arc(475, 80, 140, 155, 17);
    arc(445, 60, 325, 335, 17);
    rectangle(183, 68, 187, 70);/* showing route set for */
    rectangle(205, 87, 209, 90);/* 1B with 51R, partially */
    rectangle(192, 82, 196, 86);/* through 51T and reception plunger */
    circle(320, 86, 2);      /* on down loop line */
    setfillstyle(SOLID_FILL, YELLOW);
    floodfill(185, 69, WHITE);
    floodfill(206, 88, WHITE);
    floodfill(194, 84, WHITE);
    floodfill(320, 86, WHITE);
    rectangle(138, 68, 183, 70);
```

```cpp
      setfillstyle(SOLID_FILL, YELLOW);
      floodfill(160, 69, WHITE);/* route on 1T set */
      floodfill(175, 69, WHITE);/* route on 1AT set */
      setfillstyle(SOLID_FILL, YELLOW);/* down distant 2nd yellow */
      floodfill(86, 59, WHITE);
      setfillstyle(SOLID_FILL, BLACK);/* turning red aspect of */
      floodfill(154, 59, WHITE);/* down home off */
      setfillstyle(SOLID_FILL, YELLOW);/* set down home yellow */
      floodfill(160, 59, WHITE);
      setfillstyle(SOLID_FILL, YELLOW);/* set down home route indicator */
      floodfill(170, 60, WHITE);/* yellow */
  }
    else
    {
      cout<<"\n conditions not present for route 1B \n";
    }
};
void setroute2E()
{
    rectangle(473, 36, 477, 39);
    setfillstyle(SOLID_FILL, CYAN);/* switch 2 turned right */
    floodfill(467, 38, WHITE);/* and knob E pressed */
    floodfill(474, 37, WHITE);/* shown in cyan color */
    floodfill(498, 68, WHITE);
    floodfill(502, 72, WHITE);
    if(point53==0 && pointlock53==0 && trackcct53T==0 && trackcct2T==0 &&  LCP==0)
    {
      pointlock53=1;
      rectangle(444, 68, 495, 70);/* the path for route 2E up to */
      rectangle(505, 68, 560, 70);/* down advanced starter set & shown in */
      setfillstyle(SOLID_FILL, YELLOW);/* yellow through track circuits */
      floodfill(450, 69, WHITE);
      floodfill(480, 69, WHITE);
   floodfill(550, 69, WHITE);
      setfillstyle(SOLID_FILL, BLACK);
      floodfill(464, 59, WHITE);
      setfillstyle(SOLID_FILL, YELLOW);/* down main starter turned to */
      floodfill(470, 59, WHITE);/* yellow */
      if(DG4==1)
      {
        setfillstyle(SOLID_FILL, BLACK);
        floodfill(470, 59, WHITE);/* yellow of down main starter blanked */
        setfillstyle(SOLID_FILL, GREEN);/* and green activated when down*/
        floodfill(476, 59, WHITE);/* advanced starter turns on */
        DG2=1;/* to facilitate activation of green aspect of dn home 1 */
      }
```

```
        }
     else
     {
        cout<<"\n conditions not present for route 2E \n";
     }
};
void setroute3E()
{
     if(pointlock53==0 && trackcct53T==0 && trackcct2T==0 && LCP==0)
     {
        rectangle(453, 106, 457, 109);
        setfillstyle(SOLID_FILL, CYAN);/* switch 3 turned right and */
        floodfill(447, 108, WHITE);    /* knob E pressed */
        floodfill(454, 107, WHITE);    /* shown in cyan color */
        floodfill(498, 68, WHITE);
        floodfill(502, 72, WHITE);
        point53=1;
        pointlock53=1;
        arc(435, 70, 285, 306, 22);/* X'over 53 shown reversed */
        arc(475, 80, 140, 155, 17);/* at either end */
        arc(445, 60, 325, 335, 17);
        rectangle(464, 68, 495, 70);/*track circuit portions enclosed in */
        rectangle(505, 68, 560, 70);/* rectangles and filled with yellow */
        rectangle(457, 84, 461, 76);/* to show the path of route setting */
        setfillstyle(SOLID_FILL, YELLOW);
        floodfill(480, 69, WHITE);
        floodfill(540, 69, WHITE);
        floodfill(460, 80, WHITE);
        setfillstyle(SOLID_FILL, BLACK);/* red aspect blanked and */
        floodfill(462, 96, WHITE);/* yellow aspect activated */
        setfillstyle(SOLID_FILL, YELLOW);
        floodfill(468, 96, WHITE);
     }
        else
        {
          cout<<"\n conditions not present for route 3E \n";
        }
};
void setroute4F()
 {
  rectangle(603, 36, 607, 39);
    setfillstyle(SOLID_FILL, CYAN);/* switch 4 turned right */
    floodfill(597, 38, WHITE);      /* and knob F pressed shown */
    floodfill(604, 37, WHITE);      /* in cyan color */
    floodfill(600, 68, WHITE);
    floodfill(600, 72, WHITE);
```

```c
    BL=1; /* block instrument cleared */
    if(trackcct4T==0 && BL==1)
    {
        setfillstyle(SOLID_FILL, GREEN);/* activating the green aspect */
        floodfill(572, 53, WHITE);  /* of down advanced starter */
        floodfill(590, 59, WHITE);  /* and block instrument */
        setfillstyle(SOLID_FILL, BLACK);/* blanking red of down */
        floodfill(584, 59, WHITE);     /* advanced starter */
        DG4=1;/* to facilitate more permissive aspects of */
            /* signals in the rear */
    }
        else
        {
            cout<<"\n conditions not present for route 4F \n";
        }
};
void abnorm1()
{
  setroute1A();
  delay(5000);
  setroute1B();
};
void abnorm2()
{
  setroute1B();
  delay(5000);
  setroute1A();
};
void dnmainbklock()
{
  setroute1A();
  delay(5000);
  trackcct1T=1;
  setfillstyle(SOLID_FILL, RED);
  floodfill(160, 69, WHITE);/* track circuit 1T occupied-shown red */
  setfillstyle(SOLID_FILL, RED);
  floodfill(154, 59, WHITE);/* down home turned to red */
  setfillstyle(SOLID_FILL, BLACK);
  floodfill(160, 59, WHITE);/* down home yellow blanked */
  floodfill(80, 59, WHITE);/* green of down distant blanked */
  setfillstyle(SOLID_FILL, YELLOW);/* 1st yellow of down distant */
  floodfill(74, 59, WHITE);     /* restored */
  delay(2000);
  trackcct1AT=1;
  setfillstyle(SOLID_FILL, RED);
  floodfill(175, 69, WHITE);/* track circuit 1AT shown occupied */
```

```
    delay(2000);
trackcct51T=1;
    setfillstyle(SOLID_FILL, RED);
    floodfill(200, 69, WHITE);/* track circuit 51T shown occupied */
    delay(2000);
    trackcct01T=1;
    setfillstyle(SOLID_FILL, RED);
    floodfill(320, 69, WHITE);/* track circuit 01T shown occupied */
    delay(2000);
    trackcct01AT=1;
    setfillstyle(SOLID_FILL, RED);
    floodfill(327, 69, WHITE);/* track circuit 01AT shown occupied */
    floodfill(400, 69, WHITE);/* avoiding knob A */
    delay(500);
    setfillstyle(SOLID_FILL, BLACK);
    floodfill(160, 69, WHITE);/* back locking on tracks 1T, 1AT, 51T */
    floodfill(175, 69, WHITE);/* cleared */
    floodfill(200, 69, WHITE);
    trackcct1T=0;
    trackcct1AT=0;
    trackcct51T=0;
    setfillstyle(SOLID_FILL, BLACK);/* turning switch 1 to normal */
    floodfill(157, 38, WHITE);/* and knob A indication to normal */
    floodfill(164, 37, WHITE);
    floodfill(330, 67, WHITE);
    floodfill(335, 71, WHITE);
    floodfill(330, 69, WHITE);
    outtextxy(163, 35,"X");
};
void dnmainapplock()
 {
    setroute1A();
    delay(4000);
    setfillstyle(SOLID_FILL, BLACK);/* turning switch 1 to normal */
    floodfill(157, 38, WHITE);      /* and knob A indication */
    floodfill(164, 37, WHITE);       /* to normal */
    floodfill(330, 67, WHITE);
    floodfill(335, 71, WHITE);
    floodfill(330, 69, WHITE);
    outtextxy(163, 35,"X");
    setfillstyle(SOLID_FILL, RED);/* setting down main home signal */
    floodfill(154, 59, WHITE);    /* back to red */
    setfillstyle(SOLID_FILL, BLACK);/* blanking down main home */
    floodfill(160, 59, WHITE);/*yellow and green of down distant signal*/
    floodfill(80, 59, WHITE);
    setfillstyle(SOLID_FILL, YELLOW);/* down distant reverted */
```

```
        floodfill(74, 59, WHITE);        /* back to yellow */
        setfillstyle(SOLID_FILL, BLACK);/* blanking track circuit */
        floodfill(160, 69, WHITE);   /* indications on the whole route */
        floodfill(175, 69, WHITE);
        floodfill(200, 69, WHITE);
        floodfill(320, 69, WHITE);
        floodfill(327, 69, WHITE);
        floodfill(400, 69, WHITE);
        delay(2000);
        setroute1B();
        delay(120000);
        pointlock51=0;
        pointlock53=0;
        setroute1B();
        outtextxy(20, 160,"conditions altered after 120 secs");
    };
void mlthrecep()
    {
        setroute4F();/* train to pass through the */
        setroute2E();/* down main line with all signals */
        setroute1A();/* turning green */
    };
/* end of the program */
```

Control Table Generator Program (Ruby)

```ruby
class ControlTableGenerator

  # Scope: Public
  # Type: Instance method
  # Parameters:
  #   1. file_name - Name of input file
  #   2. graph     - Instance of graph which has whole of input data
  #   3. paths     - Array of path instances
  # Returns: Array of control table entries
  #
  # Method to compute the state of points for all given paths
  def generate_control_table file_name, graph, paths
    control_table_entries = []
    puts "Generating control table for the input file '#{file_name}'"

    # Get the required attributes from graph
    signals = graph.get_attribute :signals
    vertices = graph.get_attribute :vertices
    tracks = graph.get_attribute :tracks
    terminals = graph.get_attribute :terminals
    labels = graph.get_attribute :labels

    # Iterate through the paths to compute the point statuses
    paths.each do |path|
      # Create a ControlTable instance and set the attributes appropriately
      control_table = ControlTable.new
      control_table.set_source_vertex = path.get_source_vertex
      control_table.set_destination_vertex = path.get_destination_vertex
      control_table.set_source_signal = path.get_source_signal
      control_table.set_destination_label = path.get_destination_label
      control_table.set_controlled_by_tracks_hash = path.get_path

      signal = path.get_source_signal

      puts "  Generating control table entry for #{control_table.to_s}"
      lnd_points = case signal.get_type
          when TrainSignal::Type::HOME
            # Compute Locks and detects points field for home signal type
            compute_lnd_points_home path, vertices
          when TrainSignal::Type::CALLING_ON_HOME
            # Compute Locks and detects points field for calling on home
            # signal type
            dest_label = path.get_destination_label

            # As path was not computed for calling on home signal type
            # in the previous module, it is computed here to be used for
            # internal computations
            routes_gen = RouteGenerator.new
            routes_gen.set_attributes path.get_source_signal.get_direction, dest_label
            src_vertex = routes_gen.get_next_vertex path.get_source_signal.get_vertex
            control_table.set_source_vertex = src_vertex
            computed_path = routes_gen.get_calling_on_home_path src_vertex
            dest_vertex = (computed_path.values - computed_path.keys).first
            dest_vertex = vertices[dest_vertex]
            control_table.set_destination_vertex = dest_vertex
            # The computed path is stored as hash only and not as array
            # since array is what gets displayed in the final output
            # and hash is used in internal computations
            control_table.set_controlled_by_tracks_hash = computed_path
            compute_lnd_points_calling_on_home src_vertex, computed_path, vertices
          when TrainSignal::Type::STARTER
            # Compute Locks and detects points field for starter signal type
```

```ruby
          compute_lnd_points_starter path, vertices
        when TrainSignal::Type::SHUNT
          # Compute Locks and detects points field for shunt signal type
          compute_lnd_points_shunt_n_calling_on_home path, vertices
        when TrainSignal::Type::ADVANCED_STARTER
          # No need to compute points statuses for advanced starter signal
          # type
          {}
        end
    if signal.get_type != TrainSignal::Type::CALLING_ON_HOME
      # Convert path from hash to array for all signals except
      # calling on home
      path_hash = control_table.get_controlled_by_tracks_hash
      path_array = ControlTableGenerator.compute_path_as_array path_hash
      control_table.set_controlled_by_tracks_array = path_array
    end
    # Set the attribute for point statuses and add it to the array
    control_table.set_locks_and_detects_points = lnd_points

    puts "  Generating control table entry for #{control_table.to_s} done!"
    control_table_entries << control_table
  end

  points_entry = {}
  # Iterate through every vertex which is a point and
  # add those vertices which are point to its ID as hash
  vertices.each do |v_id, vertex|
    if vertex.is_a_point?
      if points_entry[vertex.get_point].nil?
        points_entry[vertex.get_point] = [vertex.get_data]
      elsif points_entry[vertex.get_point] &&
        !points_entry[vertex.get_point].include?(vertex.get_data)

        points_entry[vertex.get_point] << vertex.get_data
      end
    end
  end
  # Make a ControlTable instance for every state of a point in the input
  # and add the vertices that have it as point to the controlled by tracks
  # attribute
  points_entry.each do |point, point_vertices|
    (1..2).each do |n|
      control_table = ControlTable.new
      control_table.set_point_info = true
      control_table.set_source_signal = point
      control_table.set_destination_label = Vertex::PointStatus::MAPPING_SHORT[n % 2 != 0]
      puts "  Generating control table entry for #{control_table.to_s}"
      control_table.set_controlled_by_tracks_array = point_vertices
      control_table.set_locks_and_detects_points = {}
      control_table_entries << control_table
      puts "  Generating control table entry for #{control_table.to_s} done!"
    end
  end

  puts "Generating control table for the input file '#{file_name}' done"
  return control_table_entries
end


# Scope: Public
# Type: Instance method
# Parameters:
#   1. path_hash - Path as hash
# Returns: Path as array
#
# Method to convert path as hash to array
def self.compute_path_as_array path_hash
```

```ruby
    (path_hash.keys + path_hash.values).uniq
end


# Scope: Public
# Type: Instance method
# Parameters:
#   1. path      - Instance of path
#   2. vertices - Hash with vertices ID as key and Vertex instance as value
# Returns: Point statuses as hash
#
# Method to compute point statuses for extra path in case of
# shunt and calling on home signal type
def compute_shunt_n_calling_on_home_xtra_lnd_points path, vertices
    compute_lnd_points_shunt_n_calling_on_home path, vertices
end


private

    # Scope: Private
    # Type: Instance method
    # Parameters:
    #   1. path      - Path instance for a route
    #   2. vertices - Hash with vertices ID as key and Vertex instance as value
    # Returns: Locks and detects points for home signal type as hash
    #
    # Method to compute the locks and detects points for home signal type route
    def compute_lnd_points_home path, vertices
        lnd_points = {}
        # Get the source and destination vertex of route
        vertex = path.get_source_vertex.get_data
        dest_vertex = path.get_destination_vertex.get_data
        prev_vertex_obj = nil
        vertex_obj = nil
        route_path = path.get_path
        # Traverse through the computed path until the destination vertex is
        # reached
        while vertex != dest_vertex
            # Get previous Vertex instance
            prev_vertex_obj = vertex_obj
            # Get the next vertex ID to traverse and its instance
            vertex = route_path[vertex]
            vertex_obj = vertices[vertex]
            if prev_vertex_obj
                # If previous Vertex instance is not nil
                # compute points' statuses
                if prev_vertex_obj.get_track.get_data != vertex_obj.get_track.get_data &&
                    prev_vertex_obj.get_point == vertex_obj.get_point

                    # If previous vertex and vertex are not on same track
                    # and previous vertex and vertex have same point ID
                    # the point is in reverse(R)
                    lnd_points.merge!({vertex_obj.get_point => Vertex::PointStatus::REVERSE})
                elsif prev_vertex_obj.get_track.get_data == vertex_obj.get_track.get_data &&
                    prev_vertex_obj.is_a_point? && lnd_points[prev_vertex_obj.get_point].nil?

                    # If previous vertex and vertex are on same track
                    # and previous vertex is a point
                    # and previous vertex's point has not been set previously
                    # the point is in normal(N)
                    lnd_points.merge!({prev_vertex_obj.get_point => Vertex::PointStatus::NORMAL})
                end
            end
        end
        if vertex_obj.is_a_point? && lnd_points[vertex_obj.get_point].nil?
            # If vertex is a point and its point ID has been set previously
            # set the point as normal(N)
```

```ruby
      lnd_points.merge!({vertex_obj.get_point => Vertex::PointStatus::NORMAL})
    end
    return lnd_points
end


# Scope: Private
# Type: Instance method
# Parameters:
#   1. src_vertex - Source Vertex instance
#   2. path       - Path instance for a route
#   3. vertices   - Hash with vertices ID as key and Vertex instance
# Returns: Locks and detects points for calling on home signal type as hash
#
# Method to compute the locks and detects points for calling on home
# signal type
def compute_lnd_points_calling_on_home src_vertex, path, vertices
  lnd_points = {}
  # Get the source vertex ID of route
  vertex = src_vertex.get_data
  vertex_obj = nil
  prev_vertex_obj = nil
  # Traverse through the computed path till there is a path
  # In Ruby, value of a key which is not present in a hash will be nil
  while path[vertex]
    # Get the previous Vertex instance
    prev_vertex_obj = vertex_obj
    # Get the next vertex ID to traverse and its instance
    vertex = path[vertex]
    vertex_obj = vertices[vertex]
    if prev_vertex_obj
      # If previous vertex is not nil
      # compute points' statuses
      if prev_vertex_obj.get_track.get_data != vertex_obj.get_track.get_data &&
        prev_vertex_obj.get_point == vertex_obj.get_point

        # If previous vertex and vertex are not on same track
        # and previous vertex and vertex have same point ID
        # the point is in reverse(R)
        lnd_points.merge!({vertex_obj.get_point => Vertex::PointStatus::REVERSE})
      elsif prev_vertex_obj.get_track.get_data == vertex_obj.get_track.get_data &&
        prev_vertex_obj.is_a_point? && lnd_points[prev_vertex_obj.get_point].nil?

        # If previous vertex and vertex are on same track
        # and previous vertex is a point
        # and previous vertex's point has not been set previously
        # the point is in normal(N)
        lnd_points.merge!({prev_vertex_obj.get_point => Vertex::PointStatus::NORMAL})
      end
    end
  end
  return lnd_points
end


# Scope: Private
# Type: Instance method
# Parameters:
#   1. src_vertex - Source Vertex instance
#   2. path       - Path instance for a route
#   3. vertices   - Hash with vertices ID as key and Vertex instance
# Returns: Locks and detects points for starter signal type as hash
#
# Method to compute the locks and detects points for starter signal type
def compute_lnd_points_starter path, vertices
  lnd_points = {}
  # Get the source and destination vertex ID
  vertex = path.get_source_vertex.get_data
```

```
    dest_vertex = path.get_destination_vertex.get_data
    prev_vertex_obj = nil
    vertex_obj = vertices[vertex]
    route_path = path.get_path
    # Traverse through the computed path until the destination vertex is
    # reached
    while vertex != dest_vertex
      # Get the previous vertex instance
      prev_vertex_obj = vertex_obj
      # Get the next vertex ID to traverse and its instance
      vertex = route_path[vertex]
      vertex_obj = vertices[vertex]
      if prev_vertex_obj
        # If previous vertex is not nil
        # compute points' statuses
        if prev_vertex_obj.get_track.get_data != vertex_obj.get_track.get_data &&
          prev_vertex_obj.get_point == vertex_obj.get_point

          # If previous vertex and vertex are not on same track
          # and previous vertex and vertex have same point ID
          # the point is in reverse(R)
          lnd_points.merge!({vertex_obj.get_point => Vertex::PointStatus::REVERSE})
        elsif prev_vertex_obj.get_track.get_data == vertex_obj.get_track.get_data &&
          prev_vertex_obj.is_a_point? && lnd_points[prev_vertex_obj.get_point].nil?

          # If previous vertex and vertex are on same track
          # and previous vertex is a point
          # and previous vertex's point has not been set previously
          # the point is in normal(N)
          lnd_points.merge!({prev_vertex_obj.get_point => Vertex::PointStatus::NORMAL})
        end
      end
    end
    return lnd_points
end


# Scope: Private
# Type: Instance method
# Parameters:
#   1. src_vertex - Source Vertex instance
#   2. path       - Path instance for a route
#   3. vertices   - Hash with vertices ID as key and Vertex instance
# Returns: Locks and detects points for shunt and calling on home signal
# type as hash
#
# Method to compute the locks and detects points for shunt and
# calling on home signal type as hash
# Used for computing with extra path in fourth module
def compute_lnd_points_shunt_n_calling_on_home path, vertices
  lnd_points = {}
  vertex = path.get_source_vertex.get_data
  dest_vertex = path.get_destination_vertex.get_data
  prev_vertex_obj = nil
  vertex_obj = vertices[vertex]
  route_path = path.get_path
  while vertex != dest_vertex
    # Get the previous vertex instance
    prev_vertex_obj = vertex_obj
    # Get the next vertex ID to traverse and its instance
    vertex = route_path[vertex]
    vertex_obj = vertices[vertex]
    if prev_vertex_obj
      # If previous vertex is not nil
      # compute points' statuses
      if prev_vertex_obj.get_track.get_data != vertex_obj.get_track.get_data &&
        prev_vertex_obj.get_point == vertex_obj.get_point
```

```ruby
      # If previous vertex and vertex are not on same track
      # and previous vertex and vertex have same point ID
      # the point is in reverse(R)
      lnd_points.merge!({vertex_obj.get_point => Vertex::PointStatus::REVERSE})
    elsif prev_vertex_obj.get_track.get_data == vertex_obj.get_track.get_data &&
      prev_vertex_obj.is_a_point? && lnd_points[prev_vertex_obj.get_point].nil?

      # If previous vertex and vertex are on same track
      # and previous vertex is a point
      # and previous vertex's point has not been set previously
      # the point is in normal(N)
      lnd_points.merge!({prev_vertex_obj.get_point => Vertex::PointStatus::NORMAL})
      end
    end
  end
  if vertex_obj.is_a_point? && lnd_points[vertex_obj.get_point].nil?
    lnd_points.merge!({vertex_obj.get_point => Vertex::PointStatus::NORMAL})
  end
  return lnd_points
end


# Scope: Private
# Type: Instance method
# Parameters:
#   1. label  - Label ID
#   2. labels - Hash instance with label ID as key and Vertex instance as
#                 value
# Returns: Boolean value denoting if the label is not for platform
#
# Method to check if a label is not for platform
def is_non_platform_label? label, labels
  # Get the Vertex instance of the label ID
  vertex = labels[label]
  # Get the Label instance for the label ID from the Vertex instance
  v_label = vertex.get_label(label)
  !v_label.get_platform
  end
end
```

APPENDIX E (PART I)

SAFETY AND RELIABILITY REQUIREMENT OF ELECTRONIC SIGNALLING EQUIPMENT

SPECIFICATION No. RDSO/SPN/144/2014

**SIGNAL DIRECTORATE
RESEARCH DESIGN & STANDARDS
ORGANISATION
LUCKNOW – 226 011**

DOCUMENT DATA SHEET			
Designation **RDSO/SPN/144/2014**			**Revision** **3.0**
Title of Document **SAFETY AND RELIABILITY REQUIREMENT OF ELECTRONIC SIGNALLING EQUIPMENT**			
Authors: Mudit Anand **Designation: Joint Director/Signal/RDSO**			
Approved by **Name: Shri Mahesh Mangal** **Designation: Sr. Executive Director/Signal, RDSO**			
Abstract **This document defines Safety and Reliability Requirement of Electronic Signalling Equipment.**			

DOCUMENT CONTROL SHEET

NAME	ORGANIZATION	FUNCTION	LEVEL
Mudit Anand	RDSO	Member	Prepare
Mahesh Mangal	RDSO		Approve

AMENDMENTS

Version	Chapter/Annexure	Amendment	Effective date
RDSO/SPN 144/94		FIRST ISSUE	1994
RDSO/SPN 144/2004		Revision 1	25.05.2004
RDSO/SPN 144/2006		Revision 2	14.03.2006
RDSO/SPN 144/2014		Revision 3	. . 2014

Revision 3 (Details)

SN	Clause	Type of Amendment
1.	0.2	Modified
2.	0.3	Modified
3.	0.4	Modified
4.	1.1	Modified
5.	2.2, 2.4, 2.5	Modified
6.	2.7.1, 2.7.2, 2.7.3	Modified
7.	2.7.6	Added
8.	2.8.1, 2.8.2, 2.8.4, 2.8.5	Modified
9.	3.0, 3, 1, 3.4	Modified
10.	4.1, 4.4, 4.7, 4.8, 4.9	Modified
11.	5.1.1, 5.1.2, 5.1.3, 5.1.4	Modified
12.	5.2	Modified
13.	5.3.4	Modified
14.	5.4.2	Deleted
15.	5.4.3, 5.4.4	Modified
16.	6.0, 6.1, 6.2, 6.6, 6.10.6.11, 6.12,	Modified
17.	7.0, 7.2	Modified
18.	8.0, 8.1	Modified
19.	9.3	Modified
20.	9.3(1), 9.3(2), 9.3(3), 9.3(4), 9.3(5), 9.3(6), 9.3(7), 9.3(10), 9.3(11), 9.3(12)	Modified
21.	9.4,	Modified
22.	9.4.1, 9.4.2, 9.4.3, 9.4.4	Added
23.	9.7	Added
24.	10.2, 10.3, 10.4	Modified
25.	11.2, 11.3, 11.4, 11.5.1, 11.5.2	Modified
26.	12.0, 12.3, 12.4	Modified
27.	13.1, 13.6	Modified
28.	15	Deleted
29.	Annexure I	Added

GOVERNMENT OF INDIA
MINISTRY OF RAILWAYS
(RAILWAY BOARD)

INDIAN RAILWAY
STANDARD SPECIFICATION
FOR
SAFETY AND RELIABILITY REQUIREMENT OF
ELECTRONIC SIGNALLING EQUIPMENT

Serial No. RDSO/SPN/144/2014

0.0 FOREWORD

0.1 This specification is issued under the fixed serial no. RDSO/SPN/144 followed by the year of original adoption as standard or in the case of revision, the year of last revision.

0.2 This specification requires reference to the following Indian Railway Standards specifications (IRS), Indian Standards Specifications (IS), European Committee for Electrotechnical Standardization (CENELEC) and International Electrotechnical Commission(IEC)

   (i)   IRS: S 96 for DC-DC converters.

   (ii)   IRS: S 88 for Low maintenance Lead Acid Batteries.

   (iii) IRS: S 93 for Valve Regulated Lead Acid Batteries.

   (iv)   IRS: S 86 for battery chargers.

   (v)   IS: 9000 for Basic Environmental Testing procedure for electronic and electrical items.

   (vi)   IS: 9001 Guidance for Environmental Testing.

   (vii) IEC: 60034 for Degrees of protection.

   (viii) IEC: 60947/7/1; terminal blocks for copper conductors.

   (ix)   IEC: 60571: Railway Applications-Electronic Equipment used on rolling stock

   (x)   IEC: 61643; Low-voltage surge protective devices

   (xi)   IEC: 62305; Protection against lightning

(xii)   IEC: 62236;Railway Applications-Electromagnetic compatibility (Part 1 to 5)

(xiii)   IEC:62497;Railway Applications-Insulation Co-ordination (Part 1 to 2)

(xiv)   IEC:62498;Railway Applications-Environmental conditions for equipments (part 1 to 3)

(xv)   IEC:62278;Railway Applications-The Specification And Demonstration of Reliability, Availability, Maintainability and Safety (RAMS)

(xvi)   IEC: 62279; Railway Applications-Communication, signalling and processing systems-Software for railway control and protection systems.

(xvii)   IEC:62425;Railway Applications-Communication, signalling and processing systems-Safety related electronics systems for Signalling

(xviii) IEC:62280; Railway Applications-Communication, signalling and processing systems-Safety related communication(Part 1 and 2)

(xix)   IEC:62427;Railway Applications-Compatibility between rolling stock and train detection system

(xx)   IEC 61000;Testing and measurement techniques

(xxi)   IEC:60255; Measuring relays and protection equipments

(xxii)   IEC: 60352:5; Press-in connections – General requirements, test methods and practical guidance.

(xxiii) IEC: 60130; Connectors.

(xxiv)   IEC- 61508: Standard for Functional Safety of Electrical/Electronic/Programmable Electronic Safety Related Systems.

(xxv)   EN 50126: Railway applications-specification and demonstration of reliability, availability, maintainability and safety.

(xxvi)   EN 50128: Railway applications-signaling and communication-Software for Railway control and protection system.

(xxvii) EN 50129: Railway applications-Safety related electronic systems for signaling.

(xxviii) EN 50159–1 & 2: Railway applications-Signaling and Communication Safety related communication in closed and open transmission system.

0.3   Whenever in this specification, any of the above mentioned specifications are referred to by number without mentioning the year of issue, the latest issue of that specification is implied, otherwise particular issue referred to is meant.

0.4   This specification is intended chiefly to cover only the technical provisions and does not include all the necessary provisions of a contract.

1.0 SCOPE

1.1 This specification covers the reliability and safety requirements of electronic (including microprocessor/micro-controller/processor based) fail safe signalling equipments like Axle Counters, AFTCs, Electronic Interlocking Equipments, SSBPAC, UFSBI, TCAS, TPWS, IPS, DC-DC converters, LED signals, Power supply equipments, Telecom eqipments etc.

1.2 This specification shall be read with the main specification of the equipment.

1.3 Any special requirement, specified in the main specification of the equipment, shall override the requirements laid down in this specification.

2.0 GENERAL

2.1 The equipment shall be manufactured as per best engineering practices.

2.2 The cabinet shall be powder coated and shall have good aesthetic appearance. It shall conform to IP-31 class of protection as specified in IEC:60034.

2.3 The power portion of the equipment shall be clearly isolated and protected to prevent accidental contact.

2.4 All non-current carrying metals parts including shields and screens shall be bonded together and earthed. An earth terminal suitable for taking up to 4 mm diameter copper wire shall be provided. The earth terminal shall be indicated by letter 'E'. Value of earth resistance shall not be more than 1 ohm unless otherwise specified.

2.5 Outsourcing, if any, of any sub-modules or PCB shall be indicated in the 'Quality Assurance Plan' and approval of RDSO, Lucknow shall be obtained. Outsourcing of safety related sub-modules or PCBs shall be from ISO-9001 or ISO-9002 certified manufacturers only.

2.6 Necessary provision shall be made in the hardware and software for modular expansion of the equipment.

2.7 Version Control:

2.7.1 For indigenous equipments, the version number of as per the equipment shall be format given below:

DXXSXXXHXX

DXX	SXXX	HXX
Basic Design	Software	Hardware
D: Design, S: Software, H: Hardware		
XX and XXX are numeric two and three digit numbers respectively.		
Thus the initial version of any equipment will be D01S001H01.		
In every case of modification/upgradation/improvement of Basic Design, DXX will increment by one.		
In every case of modification/upgradation/improvement of system Software (executive software), SXXX will increment by one.		
In every case of modification/upgradation/improvement of Hardware, HXX will increment by one.		

2.7.2 Version number shall be displayed on each PCB as per clause 6.11 of this specification. Similarly version number shall be displayed on the name plate of the equipment as per clause 12.4 of this specification.

2.7.3 The software version number shall appear on the LCD/LED display board immediately after power ON and shall be displayed for 10 seconds. The system shall display version number of the software for 10 seconds either by giving suitable command or by pressing a button.

2.7.4 Each document/manual of the manufacturer shall contain the history of the changes in version along with accompanying changes in the manual, if any.

2.7.5 Signalling equipments which do not have any embedded software shall follow the following format for version number.

DXXHXX

DXX	HXX
Basic Design Change	Hardware Change
D: Design, H: Hardware XX are numeric two digit numbers.	
Thus the initial version of any equipment will be D01H01.	
In every case of modification/upgradation/improvement of Basic Design, DXX will increment by one.	
In every case of modification/upgradation/improvement of Hardware, HXX will increment by one.	

2.7.6 For equipments proposed under Cross Approval, the details of testing and display shall be indicated.

2.8 Change of system software

2.8.1 The system software shall be stored in separate PROM to ensure that the ROM is programmed only once and it is not be possible to modify the System Software. However, application engineers shall have the facility to modify application software as and when required. The application software shall be password protected and shall be accessed by authorized person only.

2.8.2 Any supply or installation of modified/upgraded/improved system software by a firm for equipment shall only be done with prior approval of Signal Directorate of RDSO, Lucknow. While approving the upgraded/modified/improved software with new version number, RDSO shall verify the checksum of the system software as given by the manufacturer along with version number of the equipment.

2.8.3 The Director (Q.A.)/S&T will check the version number and also the checksum of new version of the software before passing the same in the acceptance test.

2.8.4 Version number and checksum of new version shall form part of approval letter, acceptance and routine tests also.

2.8.5 After the acceptance test by Director (Q.A.)/S&T, stickers with software version no. will be written with non erasable and visible marker pen on the PROMs. In case PROMS carrying new software version are to be supplied separately for replacing PROM's of already supplied equipment, these will be sealed in a proper package and stamped with RDSO's seal before the same are dispatched to the consignee for installation. The version number and the checksum will be clearly typed or written on the sealed cover and signed by the inspecting authority.

2.8.6 Firms shall supply and install only the latest approved version of the equipment and software.

3.0   REQUIREMENTS OF SIGNAL ENGINEERING MANUAL

The equipment and its accessories shall comply with relevant Para of Signal Engineering Manual pertaining to signalling circuits using electronic equipment. The extract of relevant Para's is reproduced below:

3.1 Component failure shall be self-detecting by way of causing a signal to display a most restrictive aspect.

3.2 Failure of components which are not self-detecting shall not cause any unsafe failure of the equipment. Even simultaneous failures in different components which are not self-detecting shall not cause any unsafe failure of the equipment.

3.3 All fail-safe circuits shall work on continuous energisation principle such that open circuits in wiring, relay contacts, etc., or loss of power supply shall not cause unsafe conditions.

3.4 Common return shall not be used for vital circuits. In vital circuits, the final stage shall use fail-safe signalling relays. Isolation shall be provided between the final stage fail-safe signalling relay and the electronic device preceding it. The DC power supply shall not have any galvanic connection with the coil of the final stage signalling relay.

3.5   All electronic equipment shall have a Mean Time Between Failures (MTBF) as specified in the relevant equipment specification. Duplication of components and parts of equipment or modules may be resorted to for improvement of the reliability where necessary. Where components/parts modules are duplicated, it is desirable that provision may be made for cross checking the performance of one set by the other set and vice-versa.

3.6   Due consideration shall be given to the effects of faults in fail-safe electronic equipment to allow open or short circuit or earthing conditions and variation in component values due to ageing, replacement of faulty component with new components of specified tolerance, etc. Safety shall not be impaired as a result of multi-terminal devices failing – either open circuit, short circuit or with partial short circuit between any pair of terminals or earthing.

3.7   Special care shall be taken in the design of amplifier circuit to eliminate the possibility of self-oscillation. It is desirable that loss of safety requirements is not caused, should the amplifier go into self-oscillation due to any unforeseen contingency.

3.8   Where specific frequencies are used for safety circuits, particular care shall be taken to ensure that the frequency generating equipment is producing only the desired frequency signal. Verification shall be carried out using passive tuned filters in series with each frequency source.

3.9   The physical construction of fail-safe equipment shall be designed to eliminate the possibility of external objects causing short circuits between combinations of terminals in vital circuits. This may be achieved, for example, by adequate separation of terminals and by the fitting of protective shrouds, where necessary.

3.10   For the consideration of the fail-safe feature of an electronic safety signalling device, failure of one component for all the modes of probable faults indicated in paragraph 3.6, one at a time shall be considered. If the failure of the component under examination is not self-detecting, then simultaneous failure of other associated components shall be considered.

4.0   FAIL-SAFETY REQUIREMENT

4.1   The system should generally satisfy requirements of EN 50126: Railway applications-specification and demonstration of reliability, availability, maintainability and safety, EN 50128: Railway applications- signaling and communication-Software for Railway control and protection system, EN 50129: Railway applications-Safety related electronic systems for signaling, EN 50159: Railway applications-Signaling and Communication Safety related communication in closed and open transmission system, IEC 61508: Standard for Functional Safety of Electrical/Electronic/Programmable Electronic Safety Related Systems & IEC: 62278: The Specification And Demonstration of Reliability, Availability, Maintainability and Safety (RAMS), IEC:62279: Communication, signalling and processing systems-Software for railway control and protection systems, IEC:62425;Communication, signalling and processing systems-Safety related electronics systems for Signalling and IEC:62427; Compatibility between rolling stock and train detection system.

4.2   The system shall be designed on fail safe principles. In case of any failure whether in the hardware, software or any part of the equipment, the system and the equipments controlled by it should fail on the safe side and the system should change over to a more restrictive state.

4.3   No single failure shall result in an unsafe condition i.e. the system shall be brought to a safe state as soon as a failure occurs.

4.4   It must be ensured that if a failure of equipment occurs which by itself does not result in unsafe condition, but which in combination with a second or subsequent failure could result in an unsafe condition, then the design of the equipment must be such that the first failure is detected and negated. The probability of occurrence of a second failure, while the first failure has not been detected and negated, should be negligible.

4.5   The design of the equipment shall cater for detection and restoration of system to a safer state in case of following faults, if these are likely to result in unsafe condition:

(i)   Variation in power supply beyond its tolerance limits including its momentary or prolonged failure;

(ii)   Spikes in the power supply system, stray fields caused by traction vehicles or standby diesel generator sets;

(iii)   Insertion of PCBs in wrong card slots;

(iv)   Earthing of any component or wire or a combination of such earthing faults; and

(v)   Broken wires, damaged or dirty contacts, failure of a component to energise, loss of power supply or blown fuses etc.

4.6   The equipment shall be so constructed as to prevent unauthorized access.

4.7   Whenever power of the equipment is switched on, the equipment should wait for a manual system reset before assuming normal operational mode unless otherwise as specified in relevant specification. This may or may not be applicable in Software embedded systems.

4.8   Manual reset switch, if provided, must have an non-resettable electro-mechanical counter which should be incremented every time a reset operation is performed. System reset switch must have a locking arrangement to prevent unauthorised operation.

4.9   All vital relays, including the safe shutdown relays, shall be of approved type for use in railway signalling.

5.0 HARDWARE

5.1 COMPONENT TYPES

5.1.1 ICs and other components used in the equipment shall be of such grade that these can work satisfactorily in –400 to +850 C temperature range. Capacitors used should be certified for at least +105 deg. C. Source of procurement of components shall be given in the Quality Assurance Plan. Discrete components like diodes, transistors, SCRs etc. should conform to HIREL program of CDIL or equivalent.

5.1.2 All resistors and rectifiers used shall be rated for at least double the power which is supposed to be dissipated in them. The voltage rating of the capacitor shall be at least 50% above peak value. The resistors and capacitors shall be of tolerances not more than 5%.

5.1.3 Where ICs are used, all power supplies on cards should be locally de-coupled using a capacitor with good high frequency characteristics. The value of chip decoupler ceramic capacitor shall be 0.1 to 1μF. The value of printed circuit board decoupler electrolytic capacitor shall be 10 to 100 μF and this capacitor should be placed near to the point where power supply enters PC board.

5.2 Connectors

Connectors used should generally conform to IEC: 60130. Connectors used should be chosen considering the following;

(i) Contact resistance,

(ii) Insulation between pins,

(iii) Ruggedness and resistance to vibration,

(iv) Resistance to entry of water or other contaminants,

(v) Resistance to pressure,

(vi) Reliability,

(vii) Lifetime (number of connect/disconnect operations before failure)

(viii) Ease of connecting and disconnecting.

(ix) They should be keyed to prevent insertion in the wrong orientation, connecting the wrong pins to each other, and have locking mechanisms to ensure that they are fully inserted and cannot work loose or fall out.

(x) Connectors that apply power should be designed such that certain pins make contact before others when inserted, and break first on disconnection to protects circuits.

(xi) Connector should be easy to identify visually, rapid to assemble, require only simple tooling.

(xii)   Connectors used at radio frequencies must not change the impedance of the transmission line of which they are part. A radio-frequency connector must not allow external signals into the circuit, and must prevent leakage of energy out of the circuit. At UHF and above, silver-plating of connectors should be used to reduce losses.

5.3   DIAGNOSTIC FACILITY

5.3.1   In case of microprocessor based equipment, the system shall be provided with a front-panel alpha numeric LED/LCD display unit indicating various failures. The error code should indicate type of the failure.

5.3.2   A trouble-shooting chart should be provided indicating the action required to be taken for repair of the equipment corresponding to each error code.

5.3.3   Audiovisual alarm shall be provided to indicate failure. The audio alarm should stop when acknowledged but the visual alarm should continue till the fault is rectified.

5.3.4   Equipment should have event logging facility along with networking capability to download the log from either local or a remote place. Equipment should have a port with suitable protocol to be interfaced with Datalogger equipment for diagnostic purposes.

5.4   HOUSING RACK

5.4.1   19 ″ rack mountable and 3/4/6U high cabinets made of aluminum of minimum thickness 2mm shall be used for housing the PCB cards. The cabinet shall be powder coated. The front & backsides of the cabinets shall have facility for completely locking the equipment. The rack should have provision for natural ventilation. If required, provision for forced cooling shall be made.

5.4.2   The equipment shall be housed in a rack with a transparent front panel, if required. The rack shall have provision for natural ventilation. Ventilation openings shall be louvers of less than 3mm size covered with wire mesh for protection against entry of rodents, lizards etc. The protection shall conform to IP-31 type protection as specified in specification NO. IEC:60034.

5.4.3   Rack shall be earthed as per the code of practice for earthing & bonding for signalling equipments, RDSO/SPN/197/2008.

5.4.4   The layout of the components and wiring shall be such that all parts are easily accessible for inspection, repairs and replacement.

5.4.5   The AC input portion shall be clearly isolated and protected to prevent accidental contact.

5.4.6   Dummy slots for inserting spare PCBs shall be provided if space is available in the rack.

6.0 PRINTED CIRCUIT BOARD

The PCB shall fulfil requirements of IEC-60255 and IEC 60352–5.

6.1   PCB MATERIAL: Material for the printed circuit board shall be copper clad glass epoxy of grade FR-4 or equivalent.

6.2   OUTLINE DIMENSIONS: PCB shall normally be of standard size (e.g.3/4/6U).

6.3   BOARD THICKNESS: The thickness of PCB cards and motherboard shall be as per currently available technology. There should be no deformity in the PCB cards or the motherboard due to mounting of heavy components or due to ageing effect.

6.4   TRACK WIDTH: The track width shall be 0.5 mm nominal. In no case it should be less than 0.3 mm. Lesser width for use of SMD technology may be considered.

6.5   SPACING BETWEEN TRACKS: Spacing between tracks shall be 0.5mm nominal and in no case it shall be less than 0.3 mm. Lesser spacing for use of SMD technology may be considered.

6.6   The printed circuit cards shall be specifically designed to suit the circuitry used and no extra wires or jumpers shall be used for interconnection of components on the PCB. No piggy-back PCB shall be connected to any PCB, unless otherwise specified. The components shall be soldered with wave-soldering machine. Any exception to wave-soldering machine shall have specific approval of RDSO, Lucknow.

6.7   The cards shall be provided with testing points and the corresponding voltages/waveforms shall be indicated in the fault diagnostic procedure and service manual to facilitate testing and fault tracing.

6.8   CONFORMAL COATINGS: Assembled & tested printed boards should be given a conformal coating to enable them for functioning under adverse environmental conditions. The coating material should be properly chosen to protect the assembly from the following hazards:

(a)   Humidity;

(b)   Dust and dirt;

(c)   Airborne contaminants like smoke and chemical vapours;

(d)   Conducting particles like metal clips and filings;

(e)   Accidental short circuit by dropped tools, fasteners etc.;

(f)   Abrasion damage and

(g)   Vibration and shock (to a certain extent).

6.9   The solder masks shall be applied on the solder side and component side of the card.

6.10   Following description shall be etched/screen printed on the component side of the PCB:

(i) Component outline in the proximity of the component.

(ii) Manufacturer's name.

(iii) PCB name.

(iv) Equipment name.

(v) Part number.

6.11 Following description shall be engraved/etched/screen printed on the PCB

(i) The manufacturing serial number.

(ii) Month and year of manufacture. (iii)Version number.

6.12 Printed circuit cards shall be fitted with gold plated Euro/D type, CPU Type plug in connectors with locking arrangement. Mechanical arrangements e.g. a clip or a screw to hold the PCB in inserted position shall be provided. Screws should be countersunk and held on PCB when it is pulled out. The PCB shall be mechanically polarized so that it is not possible to insert the PCB into wrong slot.

6.13 HEAT DISSIPATING COMPONENTS: All components dissipating 3W or more power shall be mounted so that its body is not in contact with the board unless a clamp, heat sink or other means are used for proper heat dissipation.

6.14 The distribution of the power supply on the cards should be such that different voltage tracks (0, 5V etc.) follow the same route as far as possible. The track of power supplies should be as thick and wide as possible.

7.0 SOFTWARE REQUIREMENTS

Software should have been developed in conformity with a software engineering standard issued by recognized standards body such as CENELEC with special relevance to safety critical applications. The system should generally satisfy requirements of EN 50128: Railway applications-signaling and communication-Software for Railway control and protection system Particular software engineering standards used shall be specified and one complete set of such standards shall be made available to RDSO.

The software of system should generally have two layers:

(a) **Executive Software or System Software:** This Executive Software shall define what the system can do and how the various parts of the system operate together. It shall include all start up and operational safety tests (including checking the Executive Software itself) that are the parts of the processor for continual assurance of safety operation. The executive Software should have been independently verified and validated. As specified in the software Engineering Standards, full documentation on Quality Assurance Program specially the Verification and Validation (V&V) procedures carried out in-house or by any independent agency, should be made available to RDSO to check their conformity to the standards. If the procedure and documentation for V & V is considered inadequate, RDSO reserves the right to get the verification and validation of software and hardware done by an independent agency at the cost of the supplier.

(b) Application Software: It shall be containing the logic that defines how the inputs and outputs for a particular station are related. This shall be station specific. It shall not be possible to modify Executive Software. However, Application engineers should have the facility to modify application software as and when required. It should be possible to prevent unauthorized access for modifying the application software through a password protection.

The checksum of application software at the time of Factory Acceptance Test (FAT) matches with the checksum at the site (SAT) if there is no modification after FAT.

7.1   Software should be written in structured format. It should be developed in such a way that it is possible to test and validate each module independently.

7.2   The software shall be written in such a manner that in case of variable data, the possibility of using incorrect data does not exist. Further, the software should check and reject –

(a)   Use of data which is obsolete or meant for some earlier state of the system, and

(b)   Corruption of the data.

7.3   As far as possible, program flow should be independent of the input data. The program should preferably execute the same sequence of instructions in each cycle.

7.4   The use of interrupts should be kept to a bare minimum.

7.5   SELF CHECK PROCEDURES

Software should include self check procedures to detect faults in the hardware. The self check should include the following procedures:

(i)   Memory containing the vital software and data should be checked periodically so that probability of corrupted software jeopardizing the safety of the equipment is minimized.

(ii)   Components of the CPU, such as general purpose registers, program counters, stack pointers, instruction register, instruction decoder, ALU, etc., should be checked periodically as far as practicable.

7.6   Self check of the associated functional hardware as required by the hardware design should be performed periodically.

7.7   Critical and non-critical software should be segregated in the memory area so that special procedures to check the program flow may be adopted during the self check process for the critical software.

8.0 TRANSMISSION OF SIGNALLING INFORMATION

In the systems requiring transmission of vital safety, the following requirements shall be fulfilled in addition to the requirements of EN 50129 & IEC 62280;

8.1 It shall be possible to transmit the safety information over commercial voice channels/ twisted pair copper cable/OFC cables through use of proper multiplexers, unless other modes of transmission are specified by the purchaser.

8.2 The transmission protocol shall ensure required integrity of safety related information irrespective of transmission medium.

8.3 The overall system design must ensure that if the transmission link becomes inactive for more than a specified period, the safety information drain (user) will assume a restrictive and fail-safe state.

8.4 For systems relying on error prevention, all transmission equipment such as filters and amplifiers must be designated to meet specified fail safety standards.

8.5 Errors introduced or not detected at a given level in the transmission system must be detected at higher levels. Error detection methods used at any level must take into account the characteristics of the lower levels.

8.6 Error detection techniques should permit the use of standard telecommunication technology, which offers much more economic solutions than the special hardware needed to implement error prevention techniques.

8.7 Error detecting coding should not form the sole means of protection of transmitted information, but should be combined with other methods such as higher level procedures and protocols, and hardware redundancy or diversity.

8.8 Forward error correcting coding should not be used unless precautions are taken at the higher level to prevent invalid corrections from being accepted at the higher level.

9.0 ENVIRONMENTAL/CLIMATIC REQUIREMENTS

9.1 The equipment shall be capable of working in non-air conditioned environment in the field.

9.2 The equipment shall be suitable for installation on AC/DC electrified and non-electrified sections. It shall be suitable in all areas including where locomotives having thyristor controlled single phase or 3-phase induction motors haul passenger or freight trains and where chopper controlled EMU stocks are operated.

9.3 The equipment shall meet the following climatic and environmental requirements unless otherwise specified in the relevant specification and as per the IS or equivalent IEC standards:

S. No	Test		Reference	Electronic Equipment				
				Indoor	Out-door		On board	
					On Track	Track side	Inside Cab	Outside Cab
1.	**Change of temp test**		IS 9000 Part XIV Sect. 2	Yes	Yes	Yes	Yes	Yes
	Low temp	–10° C ± 3° C						
	High temp	+70° C ± 2° C						
	Rate of change in temperature	1° C/min ± 0.2° C						
	Duration	7 3 hrs at each temp. – 10° C & +70° C						
	Cycle	2 (or as otherwise specified)						
	Condition	Fully functional during test						
2.	**Dry heat test**		IS:9000 Part-III Sect 3	Yes	Yes	Yes	Yes	Yes
	Temp	+70°C ± 2° C (The rate of change of Temp shall not exceed 1° C per min averaged over 5 min.)						
	Duration	16 hrs						
	Condition	Fully functional during test						
3.	**Cold test**		IS 9000 Part II Sect. 3	Yes	Yes	Yes	Yes	Yes
	Temp	–10° C ± 3° C (The rate of change of Temp shall not exceed 1° C per min averaged over 5 min.)						
	Duration	2 hours						
	Condition	Fully functional during test.						
4.	**Damp heat test (Cyclic)**		IS 9000 Part V Sect. 2 12+12 h cycle Variant 1	Yes	Yes	Yes	Yes	Yes
	Upper temp	40° C ± 2° C						
	Humidity	95% (+1%, -5%)						
	Cycles	6						

	Condition	Fully functional during one hour period towards end of any intermediate cycles and last cycle. Stabilization shall be done at 25°± 3° C						
5.	**Damp heat test (Steady state)**		IS 9000 Part IV	Yes	Yes	Yes	Yes	Yes
	Temp	40° ± 2° C						
	Humidity	93% (+2%, -3%)						
	Severity	4 days						
	Condition	Fullyfunctional during test.						
6.	**Salt mist test**		IS 9000 Part XI procedu re 3/2	Yes Proce-dure 3	Yes Proce-dure 2	Yes Proce-dure 2	Yes Proce-dure 3	Yes Proce-dure 2
	Mist + Damp heat	Procedure 2: 2 hours + 7 days Procedure 3: 2 hours + 22 hours						
	Temp	35° ± 3° C						
	Humidity	93% (+2%, -3%)						
	Cycle	Procedure 3:3 Procedure 2:4						
	Condition	After this test, electrical parameters shall be monitored in addition to physical checks.						
7.	**Dust test**		IS 9000 Part XII	Yes	Yes	Yes	Yes	Yes
	Duration	1hour						
	Temp	40° ± 3° C						
	Condition	After this test, electrical parameters shall be monitored in addition to physical checks.						
8.	**Water Immersion test**		IS 9000 Part XV Sect. 7	No	Yes	No Yes	No	No Yes
	Head of water	0.4 m						
	Duration	24 hours						
	Condition	After this test, electrical parameters shall be monitored in addition to physical checks (Ingress of water).						

9	**Driving Rain test**		IS 9000 Part XVI Test conditio n 'C'	No	Yes	Yes	No	Yes
	Water spray for 1 hour							
	Condition	After this test, electrical parameters shall be monitored in addition to physical checks.						
10	**Bump test**		IS 9000 Part VII, Sec. 2	Yes case 1	Yes case 2	Yes case 1	Yes case 2	Yes case 2
	PCBs/Modules/units in packed condition shall be subjected to bump test as under:							
	No of bumps	Case 1: 1000 ± 10 Case 2: 4000 ± 10						
	Peak acceleration	40g						
	Pulse duration	6 ms						
	No of axes	3						
	Condition	After this test, electrical parameters shall be monitored in addition to physical checks.						
11	**Shock test (to simulate the effect of shunting shock)**		IS 9000 Part VII Sec. 1	No	Yes	Yes	Yes	Yes
	The equipment in operation shall be subjected to 3 successive shocks in each direction of three mutually perpendicular axes of the specimen, (18 shocks) of such nature that the maximum acceleration is equal to 300 m/s2. The corresponding duration of the nominal pulse shall be 18 ms. At the end of the test, the assembly shall be subjected to performance test as specified in relevant specification.							
12	**Vibration test**			TEC (IPT 1001A-revised) or IS 9000 Part VIII as specified	Yes	Yes	Yes	Yes	Yes
		Up to & including 75 Kgs. weight	**Over 75 Kgs.**						
	Freq. Range	05–350 Hz	5–150 Hz						

		± 6 mm constant displacemen t or 15m/ Sec.² constant acceleration	± 6 mm constant displacemen t or 15m/ Sec.² constant acceleration						
	Amplit ude								
	No. of axes	3	3						
	No of sweep cycle	20	10						
	Total duration	105 min	105 min						
	If resonance is observed	10 min at each resonant freq.	10 min at each resonant freq.						
	Condition	After this test, electrical parameters shall be monitored in addition to physical checks.							
13.	**EnvironmentalStress Screening tests (ESS) for Printed Circuit Boards (PCB) & sub systems** *(The manufacturer shall carry out the following ESS tests on all modules on 100% basis (except bump test) during production/ testing in the sequence as follows. Suitable records shall be maintained regarding the compliance of these tests.)*				Yes	Yes	Yes	Yes	Yes

13.1	**Thermal cycling** The PCBs shall be subjected to thermal cycling as per the procedure given below. The assembled boards are to be subjected to rapid temperature cycling as mentioned below in the power off condition.		Yes	Yes	Yes	Yes	Yes
	• This temperature cycling from 0° C to 700C, ½ Hours at each temperature for 9 cycles and 1 hour at each temp. for the 10th cycle. Dwell time of 1 hour is provided for the last cycle in order to oxidize defective solder joints exposed through thermal stress.						
	70° C, ½ Hour 1 Hour Ambient 0° C, ½ Hour r • The rate of rise/fall of temp. shall be minimum 10° C per minute. • In addition to physical checks, the electrical parameters are also to be monitored after this test.						
13.2	Power cycling: The power supply modules shall be subjected to 60 ON-OFF cycles for 1 hour. The ON-OFF switch usually provided in the modules may not be used for this purpose.		Yes	Yes	Yes	Yes	Yes

9.4   Electromagnetic Interference and Electro Magnetic Compatibility

9.4.1 All equipments shall comply with IEC: 62236. To protect against the electromagnetic interference, the following should be ensured;

(i)   Shielding at card level by providing a metallic plate over the cards. The Metallic plate shall be earthed.

(ii)   Shielding at chassis/rack level.

(iii) Circuit design for minimum radiation: Any cable will receive and radiate signals, especially when it approaches a quarter wavelength, or odd multiple thereof because it forms a resonant circuit. However even when the cable does approach these lengths, electromagnetic compatibility, EMC can be a problem.

(iv) EMC filters: EMC filters should be used for lines that carry low frequency signals like Power input cables, or other lines that carry status voltages to remove any high frequency components, leaving the low frequency elements on the line that will not radiate much. EMC filters should be placed at the entry point to the unit, and should be tightly bonded to the chassis so that no signals can enter the unit and radiate into it prior to being removed by the filter.

(v) Circuit partitioning: The circuit should be segregated into EMC critical and non-critical areas. The critical or sensitive regions should be screened or and filters added as necessary at the interfaces to prevent EMI being radiated, or to protect these circuits from the effects of EMI.

(vi) Grounding: Thick wires should be used if possible, and on printed circuit boards ground planes must be used. Critical tracks must be run above the ground plane, and they should be routed so that they do not encounter any breaks in the ground plane. Sometimes it is necessary to have a slot or break in a ground plane, and if this occurs a critical track must be routed over the plane, even if it makes it slightly longer.

(vii) Screened Enclosure: Placing the unit in a conductive enclosure that is grounded will significantly improve the performance. Where cost and possibly aesthetics are important it is possible to spray the inside of cabinets with conductive paint, although the level of screening provided will not be nearly as good as if a fully conductive metal case is used. Where high levels of EMC performance are required care should be taken to choose a case where the continuity of the screen is not breached. The case should ideally be made of as few elements as possible. At each joint there will be the possibility of radiation passing through. Where joints to occur they should be as tight as possible and they should have good continuity between them.

(viii) Screened lines and cables: When lines and cables need to pass into or out of a unit, the cables can be screened to prevent any radiation of the signals being carried or pick up of external signals. However when screened cables are needed for electromagnetic compatibility EMC applications, the screen must be bonded to the equipment signal ground as soon as it enters the unit, otherwise unwanted signals may be radiated or picked up and this would compromise the EMC compliance.

9.4.2 EMI EMC test stages

In order that a product may pass its EMC compliance, the EMC testing should be undertaken at various stages of the life of the product enlisted as under.

- Development test
- Pre-compliance test

- EMC compliance test
- Production test

9.4.3 EMC test types

Following testing for EMI/EMC generally need to be conducted.

- Conducted emissions
- Radiated emissions
- Conducted immunity
- Radiated immunity
- ESD immunity
- Transient immunity
- Surge immunity

9.4.4 The system operation and its safety should not be affected by EMI/EMC issues for which the following tests should be conducted on the equipment and as specified in the relevant specification of the equipment

EMI encountered in 25 KV AC electrified areas.

For those outdoor equipment which are used in 25 KV AC electrified areas and whose working is susceptible to the effect of electrostatic and electromagnetic induction, the following tests 9.4.4.1 and 9.4.4.2 may be performed as given in relevant specification.

9.4.4.1 One sample of the test equipment shall be subjected to static discharge test as per IEC 61000–4–2. 8 KV test voltage is to be taken, unless otherwise specified in the relevant specification. Methodology of test is given below:

(a) The equipment shall be functional and the chassis of the equipment shall be firmly grounded.

(b) A charged capacitor of 78 KV should be discharged by touching the chassis by testing probe through 330Ω resistance and 150 pF capacitor.

(c) The above discharge test should be repeated minimum 3 times.

(d) After completion of the test, the equipment shall be able to continue its normal operation.

(e) If given in the relevant specification, the discharge test should be carried out on individual card/module also.

9.4.4.2 Pantograph Interference Test: One prototype of the equipment shall be installed in the actual field condition in AC electrified traction area. An AC electric loco shall be placed in a position on the track such that distance between nearest face of the equipment and point of catenary where pantograph is touching, is about 4.0 meters.

The equipment shall be tested for its normal working during raising and lowering of the pantograph. This test will be repeated for sufficient number of times. The equipment will be tested for its normal operation after completion of the test.

9.4.4.3   Test according to IEC 61000–4–2 *(Electrostatic discharge immunity test)*

In general, the electrostatic discharge test is applicable to all equipment which is used in an environment where electrostatic discharges may occur. Direct and indirect discharges shall be considered.

9.4.4.4   Test according to IEC 61000–4–3 *(Radiated, radio-frequency, electromagnetic field immunity test)*

In general, the radiated immunity test is applicable to all products, where radio-frequency fields are present.

9.4.4.5   Test according to IEC 61000–4–4 *(Electrical fast transient/burst immunity test)*

In general, the fast transient test is applicable to products which are connected to mains or have cables (signal or control) in close proximity to mains.

9.4.4.6   Test according to IEC 61000–4–5 *(Surge immunity test)*

The surge test is applicable to products which are connected to networks leaving the building or mains in general.

9.4.4.7   Test according to IEC 61000–4–6 *(Immunity test to conducted disturbances induced by radio-frequency fields)*

In general, the conducted immunity test is applicable to products, where radio-frequency fields are present and which are connected to mains or other networks (signal or control lines).

9.4.4.8   Test according to IEC 61000–4–7 (General guide on harmonics and interharmonics measurements and instrumentation, for power supply systems and equipment connected thereto)

This technical report defines the measurement method of harmonics and interharmonics

9.4.4.9   Test according to IEC 61000–4–9 *(Pulse magnetic field immunity test)* This test is mainly applicable to products to be installed in electrical plants (for example Tele control centres in close proximity to switchgear).

9.4.4.10   Test according to IEC 61000–4–10 *(Damped oscillatory magnetic field immunity test)*

This test is mainly applicable to products to be installed in high-voltage substations.

9.4.4.11   Test according to IEC 61000–4–11 *(Voltage dips, short interruptions and voltage variations immunity test)*

This document defines the test methods to evaluate the immunity of an equipment connected to the LV system, to voltage dips, short interruptions and voltage variations. This test is applicable to equipment with a rated input current of less than 16 A per phase, connected to a.c. mains.

9.4.4.12 Test according to IEC 61000–4–14 (*Voltage fluctuation immunity test*)

In general, voltage fluctuations have an amplitude not exceeding 10 %; therefore, most equipment is not disturbed by voltage fluctuations. However, this test may be applicable to equipment intended to be installed at locations where the mains have larger fluctuations.

9.4.4.13 Test according to IEC 61000–4–17 (*Ripple on d.c. input power port immunity test*)

This test applies to equipment connected to d.c. distribution systems with external batteries charged during the operation of the equipment.

9.4.4.14 Test according to IEC 61000–4–28 (*Variation of power frequency, immunity test*)

In general, the test for variation of the power frequency is not applicable. However, it may apply to equipment intended to be installed at locations where the power frequency has large variations (for example equipment connected to an emergency power supply).

9.4.4.15 Test according to IEC 61000–4–29 (*Voltage dips, interruptions and voltage variations on d.c. input power ports, immunity tests*)

In general, this test is applicable for d.c. input power ports.

9.4.4.16 Test according to IEC 61000–4–30 (*Measurements of power quality parameters*)

This standard gives clarification on the measurement of power quality parameters.

9.5 Insulation Resistance Test: This test shall be carried out –

(a) Before the high voltage test

(b) After the high voltage test

(c) After completion of the climatic test

There shall be no appreciable change (value more than 10 Mega ohms and variation within 10%) in the values measured before and after high voltage test. After the completion of climatic test, the values shall not be less than 10 Mega ohms for the equipment at a temperature of 400 C and relative humidity 60%. The measurement shall be made at a potential of 500V DC.

9.6 Applied High Voltage Test: The equipment shall withstand for one minute without puncture and arcing a test voltage of 2000 volts rms applied between:

(a) AC line terminals and earth

(b) DC line terminals and earth

The test voltage shall be alternating of approximately sinusoidal wave form of any frequency between 50 Hz. and 100 Hz. Printed circuit cards shall be removed.

9.7 Hermetic Sealing Test

The equipment/component is to be subjected to Dust Test and Driving Rain Test as per para 9.3(7) and 9.3(9) of this specification. After the tests the equipment should not show traces of water vapour or water inside the seal, if visible and should electrically function/operate within limits. In cases where the sealed equipment/component is opaque, electrically the equipment/component should function/operate within limits.

10.0 POWER SUPPLY REQUIREMENTS

10.1 The equipment shall work on nominal voltage 24V DC (+20%, −30%) power supply or as specified in the relevant equipment specification or as approved by the purchaser.

10.2 Where separate DC-DC converters are used to derive the required DC voltages from the DC main input, these should conform to IRS: S-96 for DC-DC converters.

10.3 If the equipment has a separate Battery Charger, the battery shall be used in float charge mode from the AC mains at 230V. The battery charger shall be of low ripple voltage output type as specified in IRS: S-86 for axle counter.

10.4 A line surge suppresser (MOVRs) on input side shall be provided in the battery charger to protect against transient voltages spikes etc. For chargers meant for Telecom/Axle Counter/ EI applications, RFI, EMI filters shall also be provided both on input & output sides.

11.0 LIGHTNING AND SURGE PROTECTION FOR ELECTRONIC SIGNALLING EQUIPMENTS

11.1 The equipment shall be suitably protected against atmospheric voltage surges both for common mode (voltage that appears between phase conductors and earth) and differential mode (voltage that appears between neutral & earth) in order to limit the harmful effects of lightning.

11.2 The IEC standards 61643, 62305 pertaining to protection against lightning and surges shall apply for all electronic equipment to withstand static electricity, electric fast transient and surge voltage.

The power line of electronic signalling equipment shall have Class B & C type 2-stage protection in TT configuration. Stage 3 protection is also required for protection of power/signalling/data lines. Class B & class C type protection devices shall preferably be pluggable type to facilitate easy replacement.

11.3 11.3 Stage 1 Protection (Power line protection at Distribution Level)

(a) The Stage 1 protection shall consist of coordinated Class I/B & II/C type SPDs at the entry point of input 230V AC supply in Power/Equipment room in TT configuration in a separate wall mountable box. The Class I/B SPD shall be provided between Line to Neutral & Neutral to Earth. They shall be spark gap type voltage switching device and tested as per IEC 61643 with the following characteristics and features-

S. No.	Parameters	Limits	
		Line & Neutral	Neutral & Earth
1	Nominal Voltage (U0)	230V	230V
2	Maximum continuous operating voltage (Uc)	≥ 255V	≥ 255V
3	Lightning Impulse current (Imp) 10/350μs for each phase	≥ 25 KA,	≥ 50KA
4	Response time (Tr)	≤ 100 ñs	≤ 100 ñs
5	Voltage protection level (Up)	≤ 2.5 KV	≤ 2.5 KV
6	Short circuit withstand and follow up current extinguishing capacity without back up fuse (Isc)	≥ 3 KA	≥ 100 A
7	Operating temperature/RH	-25° C – 80° C /95%	-25° C – 80° C /95%
8	Mounted on	din rail	din rail

(b)   The Class I/B SPD will be followed by Class II/C SPD adjacent to it and connected between Line & Neutral. The device shall be a single compact varistor of proper rating and in no case a number of varistors shall be provided in parallel. It shall be voltage clamping device, thermal disconnecting type and shall be tested as per IEC 61643 with the following characteristics and features-

S. No.	Parameters	Limits
1	Nominal Voltage (U0)	230V
2	Maximum continuous operating voltage (Uc)	≥ 300V
3	Nominal discharge current between R, Y, B & N (In)	≥ 10KA, 8/20μs for each phase
4	Maximum discharge current between L & N (Imax)	≥ 40KA, 8/20μs
5	Response time (Tr)	≤ 25 ñs
6	Voltage protection level (Up) at In	≤ 1.5 KV
7	Operating temperature/RH	-25° C – 80° C/95%
8	Mounted on	din rail

(c)   Class I/B and class II/C SPDs of Stage I shall be so coordinated that the voltage protection level of the coordinated devices is ≤ 1.5 KV. As such, these devices shall be from the same manufacturer and necessary test certificate in this regard shall be submitted by the manufacturer/supplier.

11.4 Stage 2 protection (at the output side inside the distribution panel) The Stage 2 protection shall consist of Class II/C type SPDs for ≥24V– 110V AC/DC supplies at the output side inside the rack of equipment. These shall be provided for External circuits The Class II/C type SPD shall be a single compact varistor of proper rating and in no case a number of varistors shall be provided in parallel. It shall be voltage clamping device and thermal disconnecting type. They shall be tested as per IEC 61643 with the following characteristics and features-

S. No.	Parameters	Limits (between L1 & L2, L1 & E, L2 & E)	
1	Nominal Voltage (U0)	60V-110V AC/DC	24V-60V AC/DC
2	Maximum continuous operating voltage (Uc)	≥150 (AC) ≥200 (DC)	≥75 (AC) ≥100 (DC)
3	Nominal discharge current 8/20ms (In)	≥ 10KA	≥ 10KA
4	Maximum discharge current 8/20ms (Imax)	≥ 40KA	≥ 40KA
5	Response time (Tr)	≤ 25 ñs	≤ 25 ñs
6	Voltage protection level(Up)	≤1.0 KV	≤ 0.5 KV
7	Operating temperature /RH	−25° C – 80° C/95%	−25° C – 80° C/95%
8	Mounted on	Din rail	Din rail

11.5 Stage 3 protection (Protection for Power/signalling/data lines)

All external Power/signalling/data lines (AC/DC) shall be protected by using preferably pluggable stage 3 surge protection devices which consists of a combination of varistors/suppressor diodes and GD tube with voltage and current limiting facilities.

11.5.1 Power line Protection (Class D)

The device for power line protection shall be of Class D type. This shall have an indication function to indicate the prospective life and failure mode to facilitate the replacement of failed SPDs. This shall be thermal disconnecting type and equipped with potential free contact for remote monitoring. This protection shall be in compliance to IEC 61643 with following characteristics:

Nominal Voltage (U0)	24V	48V	60V	110V	230V
Max. continuous operating voltage (Uc)	30V	60V	75V	150V	253V
Rated load current (IL)	16A	16A	16A	16A	16A
Nominal discharge current (In) 8/20 µs	≥700A	≥700A	≥700A	≥2.0KA	≥2.5KA
Max discharge current (Imax) 8/20 µs	≥2KA	≥2KA	≥2KA	≥5KA	≥5KA

Voltage protection level (U$_P$)	≤200V	≤350V	≤500V	≤700V	≤1100V
Response time (T$_r$)	≤25 ñs	≤25 ñs	≤25 ñs	≤25 ñs	≤25 ñs

Note: Minor variations from above given parameters shall be acceptable.

11.5.2 Signalling/Data line protection

These devices shall preferably have an indication function to indicate the prospective life and failure mode to facilitate the replacement of failed SPDs. If the device has any component which comes in series with data/signalling lines, the module shall have "make before break" feature so that taking out of pluggable module does not disconnect the line. This protection shall be in compliance to IEC 61643 with the following characteristics:

Nominal Voltage(U0)	5V	12V	24V	48V
Arrester Rated Voltage(UC)	6V	13V	28V	50V
Rated load current(IL)	≥250mA	≥250mA	≥250mA	≥250mA
Total discharge current, 8/20 µs (In)	≥20KA	≥20KA	≥20KA	≥20KA
Lightning test current 10/350 µs	≥2.5KA	≥2.5KA	≥2.5KA	≥2.5KA
Voltage protection level (UP)	≤10V	≤18V	≤30V	≤70V

Note: Minor variations from above given parameters shall be acceptable.

11.5.3 If power supply/data/signalling lines (AC/DC) are carried through overhead wires or cables above ground to any nearby building or any location outside the equipment room, additional protection of Stage 2 (Class C) type shall be used at such locations for power supply lines and Stage 3 protection for signal/data lines.

11.6 Coordinated type Class B & C arrestor shall be provided in a separate enclosure adjacent to each other. This enclosure should be wall- mounting type.

11.7 Length of all cable connection from input supply and earth busbar to SPDs shall be minimum possible. This shall be ensured at installation time.

11.8 Stage 1 & Stage 2 (Class B & C) protection should be from the same manufacturer/ supplier. Manufacturer shall provide Stage 1 & Stage 2 protection. Stage 3 protection shall be got provided by Railways separately.

11.9 The cross sectional area of the copper conductor for first stage protection shall not be <16 mm^2 and for second stage shall not be < 10 mm^2

11.10 Batch test report of OEM should be submitted by the manufacturer/supplier of Lightning & Surge protection devices to the IPS manufacturer at the time of supply of these devices. Copy of the same shall be submitted by manufacturer to RDSO at the time of acceptance test of equipment.

12.0 MARKING

All indigenous equipments shall meet the following requirement;

12.1 All markings/indications shall be easily legible and durable. Where the marking is by use of labels, the labels shall be metallic and shall be firmly fixed and shall not be capable of being removed by hand. Durability of marking shall be checked by rubbing the marking by hand with a piece of cloth soaked with petroleum spirit. This requirement shall also be met after completion of climatic test.

12.2 All markings/indications shall be placed in the vicinity of the components to which these refer and shall not be placed on removable parts, if these parts can be replaced in such a way that the marking/indications can become misleading.

12.3 The words 'Indian Railway Property' shall be etched, engraved, screen printed or embossed on the equipment at a conspicuous position. For it, the size of the letters shall be chosen depending upon the equipment but shall not be less than 20 mm high in any case.

12.4 The following information shall be engraved on the anodized name plate firmly attached to the equipment.

(a) Name or trademark of the manufacturer.

(b) Specification number.

(c) Serial number of the equipment.

(d) Installation for which meant.

(e) Month and year of manufacture.

(f) Version number.

(g) Equipment Name

13.0 DOCUMENTATION

Two copies of the following manuals shall be supplied:

13.1 System description manual and Instruction Manual

13.2 Installation and Maintenance Manual including Dos & Don'ts.

13.3 Mechanical drawings of each sub-system/rack.

13.4 Guaranteed performance data, technical & other particulars of the equipment.

13.5 Schematic block diagram showing mounting arrangement of various components & details of each type of assembled PCB.

13.6 Trouble shooting procedures along with test voltages and waveforms at various test points in the PCBs. The possible error codes and thereafter actions required should be mentioned in the maintenance manual.

13.7 Details of software viz. Source code, algorithm, flow chart, machine code along with test/validation procedure used and the results thereof.

13.8 Details of Hardware e.g. schematic diagrams of the system circuits/components, details for each type of assembled PCB and part-list.

13.9 Pre-commissioning check list.

14.0 PACKING

14.1 The equipment and its sub assemblies shall be wrapped in bubble sheet and then packed in thermo Cole boxes and the empty spaces shall be filled with suitable filling material. All PCBs shall be enclosed in anti-static shield cover. The equipment shall be finally packed in a wooden case of sufficient strength so that it can withstand bumps and jerks encountered in a road/rail journey.

14.2 Each box shall be marked with code numbers, contents and name of manufacturer. The upside shall be indicated with an arrow. Boxes should have standard signages to indicate the correct position and precaution "Handle with Care" with necessary instructions.

14.3 Printed circuit boards shall be separately and individually packed to prevent damage.

****

Annexure I

List of Abbreviations

SN	Abbreviation	Full Form
1	AC	Alternating Current
2	AFTC	Audio Frequency Track Circuit
3	ALU	Arithmetic Logic Unit
4	CDIL	Continental Device India Limited
5	CENELAC	European Committee for Electro technical Standardization
6	CPU	Central Processing Unit
7	DC	Direct Current
8	ESD	Electrostatic Discharge
9	EMU	Electrical Multiple Unit
10	EMI	Electromagnetic Interference
11	EMC	Electromagnetic Compatibility
12	EPROM	Electrically Programmable Read Only Memory
13	FAT	Factory Acceptance Test
14	GD	Gas Discharge
15	HIREL	High Reliability
16	IC	Integrated Chip
17	IEC	International Electrotechnical Commission
18	IP	Internet Protocol
19	IPS	Integrated Power Supply
20	IRS	Indian Railway Standards
21	IS	Indian Standards
22	ISO	International Organization for Standardization
23	LCD	Liquid Crystal Display
24	LED	Light Emitting Diode
25	MTBF	Mean Time Between Failure
26	MTBWSF	Mean Time Between Wrong Side Failure
27	OEM	Original Equipment Manufacturer
28	OFC	Optical FibreCommunication
29	PCB	Printed Circuit Board
30	RDSO	Research Design and Standards Organisation
31	RFI	Radio Frequency Interference
32	PROM	Programmable Read Only Memory
33	QA	Quality Assurance
34	SAT	Site Acceptance Test
35	S&T	Signalling and Telecommunication
36	SCR	Silicon ControlledRectifier
37	SMD	Surface Mount Technology
38	SPD	Surge Protection Device
39	SSI	Solid State Interlocking
40	UHF	Ultra High Frequency
41	V&V	Verification and Validation

APPENDIX E (PART II)

ELECTRONIC INTERLOCKING

(FORMELY KNOWN AS SOLID STATE INTERLOCKING)

SPECIFICATION NO. RDSO/SPN/192/2005

**SIGNAL DIRECTORATE
RESEARCH DESIGNS & STANDARDS
ORGANISATION
MINISTRY OF RAILWAYS
MANAK NAGAR
LUCKNOW – 226 011**

DOCUMENT DATA SHEET				
Designation RDSO/ SPN/192/2005				Version 0.0
Title of Document Specification for Electronic Interlocking.				
Authors: See Document Control Sheet Signed by: Name: Anurag Goyal Designation: Jt. Director/Signal/RDSO				
Approved by Name: Shri G. D. Bhatia Designation: Sr. Executive Director/Signal, RDSO				
Abstract This document defines Electronic Interlocking				

DOCUMENT CONTROL SHEET

NAME	ORGANIZATION	FUNCTION	LEVEL
Anurag Goyal	RDSO	Member	Prepare
G.D. Bhatia	RDSO		Approve

AMENDMENTS

Version	Chapter/ Annexure	Amendment	Effective date
RDSO/SPN/192/2005		FIRST ISSUE	

Government of India
Ministry of Railways (Railway Board) INDIAN RAILWAY
STANDARD SPECIFICATION
FOR
ELECTRONIC INTERLOCKING

SERIAL NO. RDSO/SPN/192/2005

0.   FOREWORD

0.1   This specification is issued under the fixed serial No. RDSO/SPN/192/2005 followed by the year of original adoption as standard or in case of revision, the year of latest revision.

0.2   This specification requires reference to the latest version of following specifications: –

1	IRS: S 36	Relay interlocking systems
2	IRS: S23*	Electrical signalling and interlocking equipment
3	RDSO/SPN/144	Safety and reliability requirement of electronic signalling equipment.
4	IS: 9000*	Basic environmental testing procedures for electronic and electrical items.
5	IS 2147–62*	Degrees of protection provided by enclosure for low voltage switchgear and control gear.
6	ISO 9001	Quality Systems-model for quality assurance in design, development, production, installation and serving.
7	EN50126	Railway applications-specification and demonstration of reliability, availability, maintainability and safety.
8	EN50128	Railway applications-signaling and communication-Software for Railway control and protection systen.
9	EN50129	Railway applications-Safety related electronic systems for signaling.
10	EN50159–1 & 2	Railway applications-Signaling and Communication Safety related communication in closed and open transmission system.
11	IEC 529/EN 60529	Specification for degree of protection provided by enclosures (IP code).
12	EN 61000.4.2	Electromagnetic compatibility (EMC)-testing and measurement techniques-electrostatic discharge immunity test and basic EMC.
13	EN 61000.4.4	Electromagnetic compatibility – testing and measurement techniques-electrostatic fast transient/burst immunity test and basic EMC publication.
14	EN 61000.4.5	Electromagnetic compatibility-testing and measurement techniques-surge and immunity test.
15	IRS: S-99	Data Logger System.
16	RDSO/SPN/186	Domino Type Control Panel for Railway Signalling

* Or equivalent Recognized International standard. The supplier shall submit a copy of the same for verification.

Whenever, reference to any specification appears in this document, it shall be taken as a reference to the latest version of that specification.

0.3 ABBREVIATIONS

S. No.	ABBREVIATION	EXPANDED FORM
1.	ABS	AUTOMATIC BLOCK SIGNALLING
2.	ATP	AUTOMATIC TRAIN PROTECTION
3.	CA	CROSS ACCEPTANCE
4.	CCIP	CONTROL CUM INDICATION PANEL
5.	CD	COMPACT DISC
6.	CENELEC	EUROPEAN COMMITTEE FOR ELECTRO TECHNICAL STANDARDIZATION
7.	CIU	CENTRAL INTERLOCKING UNIT
8.	CMU	CENTRAL MONITORING UNIT
9.	CTC	CENTRALISED TRAIN CONTROL
10.	EI	ELECTRONIC INTERLOCKING
11.	EMU	ELECTRICAL MULTIPLE UNIT
12.	EPROM	ERASABLE PROGRAMMABLE READ ONLY MEMORY
13.	IBS	INTERMEDIATE BLOCK SIGNALLING
14.	I/O	INPUT/OUTPUT
15.	ISA	INDEPENDENT SAFETY AUDITOR
16.	MTBF	MEAN TIME BETWEEN FAILURE
17.	MTBWSF	MEAN TIME BETWEEN WRONG SIDE FAILURE
18.	MTTR	MEAN TIME TO REPAIR
19.	MT	MAINTENANCE TERMINAL
20.	OC	OBJECT CONTROLLER
21.	OFC	OPTICAL FIBRE CABLE
22.	PC	PERSONAL COMPUTER
23.	PCB	PRINTED CIRCUIT BOARD
24.	QA	QUALITY ASSURANCE
25.	QAP	QUALITY ASSURANCE PROGRAM
26.	SEM	SIGNAL ENGINEERING MANUAL
27.	SIL	SAFETY INTEGRITY LEVEL
28.	STR	SCHEDULE OF TECHNICAL REQUIREMENTS
29.	TOT	TRANSFER OF TECHNOLOGY
30.	UV	ULTRA VIOLET
31.	VDU	VISUAL DISPLAY UNIT
32.	VGA	VIDEO GRAPHIC ARRAY

1. SCOPE

1.0   This specification covers the technical requirements of Electronic Interlocking.

1.1   The EI covered in this specification shall be a microprocessor based equipment used for the operation of points, signals, level crossing gates, block working with adjacent station, releasing of crank handle for manual operation of points and other controls like slots etc. through a control cum indication panel or VDU based control panel. It shall be capable of future interfacing with ATP & CTC systems.

1.2 In case of End Cabin/Multi Cabin working, it shall be possible to interface more than one CCIP or VDU control terminal or both with the EI.

2. TERMINOLOGY

2.1 For the purpose of this specification, the terminology given in latest version of IRS: S 23 and RDSO/SPN/144 shall apply.

3. GENERAL REQUIREMENTS

3.1 The system shall provide all the interlocking, control and indication functions as per approved interlocking plan, selection table and panel diagram of the station.

3.2 The system shall have facility of monitoring of internal variables as well as status of I/O.

3.3 The system shall be suitable for working on sections having 25 kV AC traction and where passenger/freight trains hauled by single phase thyristor controlled or three phase induction motor controlled AC locomotives or chopper controlled EMU stock are operated.

3.4 The system shall be capable of working in conjunction with the control cum indication panel or a VDU or both as per clause 5.3 as required by Railways.

3.5 The system should have capability to interface with Block Working. It should also be capable of interfacing with IBS, ABS including interfacing with outlying yards and sidings. Supplier shall submit interface details.

3.6 The system shall be capable for working in non air-conditioned environment and ambient temperature range between –10° C to 70° C and Relative Humidity unto 95% at 40° C.

3.7 The system shall be provided in a dust protected cabinet. If forced cooling is required, the cooling fans shall operate on system power supply with over current protection arrangement. The failure of any one of the fans shall give an alarm to the operator.

3.8 The equipment shall be so constructed as to prevent unauthorized access to the system.

3.9 Necessary provision shall be made in the hardware and software for modular expansion of the system. For large stations, which cannot be covered by one EI, it shall be possible to connect more than one EI preferably through a serial channel. The communication channel provided between various EI shall comply with the requirements for transmission of vital safety information as laid down in relevant clause of latest version of RDSO/SPN/144/2004.

3.10 EI shall have user-friendly graphic based design tool to generate station specific application software to carry out future yard modifications.

3.11 For all vital inputs/outputs, double cutting arrangement shall be provided.

3.12 Either OFC or twisted pair cable shall be used for all vital connections.

3.13 REQUIREMENT OF ELECTRONIC INTERLOCKING

3.13.1 Both hardware & software of EI must meet SIL-4 as defined in CENELEC Standards. If the system is developed using any equivalent International standard other than CENELEC, a copy of standards followed shall be submitted with application. The certificate of validator certifying that the system is equivalent to SIL-4 compliant shall also be submitted.

3.13.2 The EI system software as well as warm/hot standby changeover software should have been independently verified and validated including its offered configuration by third party. User Railway shall verify application software pertaining to yard data.

3.13.3 The firm manufacturing EI, when applying for type approval or cross – acceptance approval shall submit documentary proof of independent validation as per CENELEC Standards or equivalent standard alongwith complete safety case.

3.13.4 The firm shall give details of all modifications carried out in the system after initial validation/ approval. Date of each modification with brief reasons for undertaking modifications shall be given. All modifications must have got approval of original validating agency/approving agency.

3.13.5 The next level Signal control circuits like Cascading of Signal aspects, Red lamp protection etc. shall be achievable through Software only.

3.13.6 The audio-visual alarm shall be available for Approach locking, Button stucking etc. in EI.

3.13.7 CIU shall have log of all counters provided at Panel like Emergency Route cancellation, Calling on signal, Emergency Point operation, Overlap release operation etc. so that in case, operation commands are given through VDU in place of CCIP, then proper working of counters shall be possible and readings of all counters can be read as and when required.

4. INTERLOCKING REQUIREMENTS

4.1 The system shall meet the interlocking requirements as specified in Cl.4.0 of IRS: S 36.

5. SYSTEM COMPOSITION

5.1 The EI system shall consist of the following:

5.1.1 Microprocessor based interlocking equipment to read the yard and panel inputs, process them in a fail-safe manner as per the selection table and generate required outputs.

5.1.2 Cycle time and response time to read and process the input shall be fast enough to ensure safety and avoid any apparent delay. Cycle time and response time of the system shall be clearly indicated.

5.1.3 Requirement of spare parts of each type for the first line maintenance shall be indicated to meet system availability with Mean time to repair (MTTR) being not more than 6 hours.

5.1.4 Domino type Control Cum Indication Panel (CCIP) with panel processor having standby processor or VDU control terminal as required by purchaser.

5.1.5 Maintenance terminal (MT) with display, keyboard, printer and event logging facility for minimum 10, 00, 000 events. The system shall have facility for automatic serial data transfer to a central monitoring unit through data logger. The protocol for this communication shall be as per Data Logger specification No. IRS: S-99.

5.1.6 Relay rack along with required number of approved type of relays or OCs.

5.1.7 OBJECT CONTROLLER

5.1.7.1 OC shall be a Processor based system having similar architecture as of CIU. It shall work as slave unit of CIU through duplicated serial communication and placed within 15 Km. radius from CIU. The OC shall drive the field gears (Points, Signals & Relays) and take feedback (Inputs) from various field gears without any modification/change in the design of outdoor Signalling equipment.

5.1.7.2 OC shall be normally placed in field locations.

5.1.7.3 The medium of communication between CIU and OCs shall be OFC provided on a ring basis. In case of communication failure between CIU & OC, all the outputs shall be brought to safe state whenever two consecutive telegrams are not received in stipulated time period.

5.1.7.4 All the inputs and outputs of OCs shall be isolated.

5.1.7.5 OC shall carry out the supervisory function to check the proper level of system voltages at critical points to ensure proper working of the system and shall also check the health of the complete system.

5.1.7.6 Occurrence of any error in any OC or hardware fault leading to unsafe condition shall immediately withdraw all output commands and remove the source supply to outputs. Functionally, each OC should be independent from other OC. Error in one OC should not affect the working of other OCs.

5.1.7.7 If the system is developed using OCs, then it shall be developed in following two phases:

(i) In first phase, the object controller shall have only Relay driver cards duly validated and meeting all safety requirements to drive Points, Signals and other field gears. The approval shall be given to provide system with this arrangement over the stations of Indian Railways.

(ii) In second phase, after successful testing & working of above mentioned system for specified time as decided by RDSO, Solid State Point and Signal lamp modules shall be developed to the satisfaction of validators and RDSO.

5.1.7.8 In case of Cross Acceptance, EI with OCs may be directly accepted provided these have performed satisfactorily for the quantity and period as specified in Cross Acceptance procedure of RDSO.

5.2   CCIP shall conform to relevant clauses of IRS: S 36 and RDSO/SPN/186. It shall be provided with push buttons/control switches for individual operation of points, clearing of signals, releasing of crank handle/ground lever frame/gate controls, cancellation of routes and other functions as covered by IRS: S 36 including block signalling, auto signal, IB signal, adjacent yard layout, to facilitate indication or operation cum indication as per requirement.

5.3   CONTROL TERMINAL WITH VDU DISPLAY:

5.3.1   If required by the purchaser, a control terminal with VDU display in lieu of or in addition to conventional CCIP shall be supplied. This will consist of:

(i) A latest PC, colour VDU monitor with minimum size of 17"(43 cm.) as specified by purchaser.

(ii) A Key Board & mouse and

(iii) Suitable interface to continuously display the current position/status of various field equipment and track circuits.

A flashing indication shall be provided on the VDU to indicate healthy condition of the main system, communication channel and panel processor.

Three dot markers in Red, Blue & Green colours respectively shall also be displayed prominently at conspicuous location on the VDU terminal to indicate that the colour monitor is healthy and all the three colours (Red, Blue & Green) are present in right proportion.

5.3.2   The control terminal shall work with 230V ± 10%, 50Hz AC power supply, for which an UPS of adequate capacity shall be supplied along with the system.

5.3.3   A colour monitor (minimum VGA or better) shall be used for the VDU of the control terminal. It shall be possible to display the complete yard layout including the section on the monitor. It shall also have facility for displaying a portion of the yard or section in an enlarged mode, if required.

5.3.4   The current position/status of various field equipments and track circuits shall be displayed on the VDU using different colours/symbols, as desired by the purchaser.

5.3.5   The system shall have suitable interface to receive and process the information for displaying the status of field equipment on the control terminal. This interface shall be of standard type like RS 232 or any other approved type.

5.3.6   Availability of communication channel shall be indicated by a constantly flashing indication. Whenever the serial channel goes faulty, a suitable error message shall be displayed on the terminal.

6.   **HARDWARE AND FAIL-SAFETY**

6.1   Requirements of SEM as laid down in relevant clause of latest version of RDSO/SPN/144 shall be complied.

6.2 COMPONENTS

Components used shall comply with relevant clause of latest version of RDSO/SPN/144 and should be commercially available.

6.3 PROTECTION AGAINST ELECTROMAGNETIC AND ELECTROSTATIC INTERFERENCE

The requirements laid down in relevant clause of latest version of RDSO/SPN/144 shall be complied. The equipment chassis shall be connected to suitable earth.

6.4 PRINTED CIRCUIT BOARD

6.4.1 The requirements laid down in relevant clause of latest version of RDSO/SPN/144 shall be complied.

6.4.2 Each card shall be marked with running serial number for identification of individual cards.

6.5 FAIL-SAFETY

6.5.1 The requirements laid down in relevant clause of latest version of RDSO/SPN/144 shall be complied.

6.5.2 Either or both of hardware and software redundancy shall be provided to ensure that any single fault does not lead to unsafe failure.

6.5.3 MTBWSF should be minimum 109 hours.

6.6 The system shall have provision for accommodating additional 25% of I/O cards.

7.0 SYSTEM ARCHITECHTURE

7.1 One of the following architectures shall be employed in the system.

(a) Single Hardware architecture with diverse software. In addition, hot/Warm standby processor(s)/system shall be provided with facility of automatic changeover.

In case of Warm standby system, the standby system should start functioning with a time delay of approximately 120 secs. of failure of main system. Preferably, the train operation shall not be affected or otherwise, there shall be no unsafe occurrence due to switching over from main system to standby system.

In case of hot standby system, train operation shall not be affected. It should also be ensured that the fault, which affected the main processor/system, does not affect the hot standby processor/system.

(b) Two out of two hardware architecture with identical hardware and identical or diverse software. In addition, warm standby/hot standby processor(s)/system using similar 2 out of 2 hardware and software architecture shall be provided with facility of automatic changeover.

In case of Warm standby system, the standby system should start functioning with a time delay of approximately 120 secs. of failure of the main system. Preferably, the train operation shall not be affected or otherwise, there shall be no unsafe occurrence due to switching over from main system to standby system.

In case of hot standby system, train operation shall not be affected. It should also be ensured that the fault, which affected the main processor/system, does not affect the hot standby processor/system.

(c) Two out of three hardware architecture with identical hardware and identical or diverse software.

7.2 MAINTENANCE AND DIAGNOSTIC AIDS

7.2.1 MT consisting of a standard PC with printer from a reputed manufacturer shall be provided for following Operations:-

(i) Display of the current status of points, signals, controls etc. of the yard.

(ii) Storage of minimum one month data or 10, 00, 000 events.

(iii) Display of recorded events and

(iv) Data transfer to floppy, CD, flash memory or any other storage media.

(v) Transfer of recorded events to external data logger.

7.2.2 Result of the failure of any card/module in the system should be clearly indicated. The supplier should also indicate process of replacing such defective cards/modules.

7.2.3 Control operation of yard functions shall not be possible from the maintenance terminal.

7.2.4 In case of any module/card becoming faulty, this fact should be displayed on MT with diagnostic facility to identify faulty module/card.

8. SOFTWARE REQUIREMENTS

8.0 The software of system should have two layers:

(a) Executive Software or System Software

This Executive Software shall define what the system can do and how the various parts of the system operate together. It shall include all start up and operational safety tests (including checking the Executive Software itself) that are the parts of the processor for continual assurance of safety operation.

(b) Application Software

It shall be containing the logic that defines how the inputs and outputs for a particular station are related. This shall be station specific.

The Executive Software and Application Software shall be programmed into Read Only Memories (ROM) by the manufacturer. Both the ROMs shall be separated & isolated from each other. It shall not be possible to modify Executive Software. However, Application engineers should have the facility to modify application software as and when required.

8.1 Software used in EI should have been developed in conformity with a software engineering standard issued by recognized standards body such as CENELEC with special relevance to safety critical applications. Particular software engineering standards used shall be specified and one complete set of such standards shall be made available to RDSO.

8.2   The selected EI Software should have been independently verified and validated. As specified in the software Engineering Standards, full documentation on Quality Assurance Program specially the Verification and Validation (V&V) procedures carried out in-house or by any independent agency, should be made available to RDSO to check their conformity to the standards. If the procedure and documentation for V & V is considered inadequate, RDSO reserves the right to get the verification and validation of software and hardware done by an independent agency at the cost of the supplier.

8.3   The system shall conform to software requirements and self-check procedures as laid down in relevant clause of latest version of RDSO/SPN/144.

8.4   SELF CHECK PROCEDURES:

8.5   Self-check of the associated functional hardware as required by the hardware design should be performed periodically as laid down in relevant clause of latest version of RDSO/ SPN/144.

Sufficient self-check should be built into the system to detect possible hardware faults.

8.5.1   Integrity of the final vital output of the system for control of the field equipment should be continuously checked by reading both front & back contacts of relays to guard against inadvertent operation of the equipment.

9.   POWER SUPPLY REQUIREMENTS

9.1   The EI shall work on 110V/60V/24V/12V DC power supply.

9.2   Two different voltages shall be used, one to drive EI equipment and the other for receiving the inputs from the field gears.

9.3   The short circuit protection shall be provided.

9.4   The required protection shall be provided to protect from any malfunctioning due to false/ spurious feed.

9.5   Suitable surge protection and proper earthling arrangement shall be provided in the power supply system to protect against transient voltages, lightning & spikes etc.

9.6   If CCIP and CIU are in separate building, then lightning and surge protection has to be provided for each core of copper cable connecting CCIP and CIU or else OFC cable shall be used to connect CCIP & CIU.

9.7   A detailed Power supply arrangement diagram/circuit shall be provided.

9.8   Power supply arrangement for individual processor should be such that, in case of fault in power supply of one processor, all processors should not cease to function

simultaneously. It should be possible to switch off and take out faulty processor for repairing/ replacement without affecting working of the balance system.

10. INFORMATION TO BE FURNISHED BY THE MANUFACTURER/SUPPLIER

The manufacturer shall supply the following information.

(a) Design approach for the system.

(b) Functions achieved in hardware & software.

(c) Mode of interaction between hardware &software.

(d) Salient feature through which fail safety has been achieved e.g. use of a watchdog timer, automatic shut down etc.

(e) Proof of safety in the form of process adopted for safety analysis and result thereof.

(f) Full documentation of Software Engineering followed during development.

(g) Full documentation of verification and validation procedure, Quality Assurance Program along with report and certificate from in-house Quality Assurance (QA) Group or an Independent Safety Auditor (ISA).

(h) If the Railways consider software validation necessary, the manufacturer/supplier will supply all the documents etc. to the Validator nominated by the Railways.

(i) Complete application software with facility for EPROM programming for entering yard data.

(j) In case of Cross-Acceptance, the firm should submit the performance feedback as given below:

- Name of System/Equipment   :

- Make   :

- Model/Version No.   :

- User Railway & Section   :

- Maximum Sectional Speed   :

- Average number of Trains per day   :

- Application of System/Equipment   :

- Problems faced and solutions evolved   :

- Failure data may be submitted as per format given below :-

Location	No. of System/ Eqpt.	Date of commis-sioning	Total hours in use	No. of safe side failures	No. of unsafe failures	MTBF	MTBWSF	MTTR
Total								

Proven ness criteria of Equipment Usage of same Type/Make & Model/Version shall be as under:-

S.No.	Category of Equipment/System	Minimum no. of Equipment	Equipment Hours in use
1.	Solid State Interlocking	25	2, 16, 000

(i)   At least 20% of the equipment/system, with a minimum of 10, should be in continuous operation for a minimum period of 720 days.

(ii)   If the offered equipment has undergone minor hardware/software up-gradation to improve functionality/safety of the equipment in recent past, then the equipment utilisation of the earlier version (prior to minor modifications) can be considered for the proveness. However, in such cases, a minimum of 10 (Ten) equipments should be in continuous operation for a minimum period of 180 days.

10.1   The manufacturer shall supply the following documentation/manuals:

(i)   Installation & Maintenance Manual with pre-commissioning check list.

(ii)   Diagnostic aids including troubleshooting charts: A trouble-shooting chart shall also be provided to indicate the step-by-step actions to be taken in case of failure of the equipment. It shall be possible to rectify the fault by replacement of defective PCB card by the maintainer at site.

(iii) Details of Hardware e.g. schematic diagrams of the system circuits/components, details for each type of assembled PCB.

(iv)   Details of software algorithm flow chart along with test/validation procedure used and the results thereof.

(v)   Version No. of Signalling equipment shall be as per RDSO/SPN/144. In case of Cross-acceptance, Version No. as per manufacturer's practice may be accepted.

(vi)   Software checksum of EPROM(s) shall be provided as per RDSO/SPN/144.

10.2   The manufacturer shall provide the following certifications from approved validation agency:

(i)   Correctness and safety of the software.

(ii)   Reliability and fail-safety of the interlocking system.

(iii) Details of modifications carried out in the system and its subsequent validation.

(iv)   Expected MTBF.

(v)   Expected MTBWSF.

(vi)   Expected MTTR.

11. TESTS AND REQUIREMENTS

11.1   Conditions of Tests

Unless otherwise specified all tests shall be carried out at ambient atmospheric conditions.

11.2   For inspection of material, relevant clauses of IRS: S 23 and RDSO/SPN/144 shall apply.

11.2.1 TEST EQUIPMENT:

The firm should have all essential Testing Equipments as per latest STR.

11.3   TYPE TESTS:

11.3.1 Standard RDSO layout shall be used for conducting type tests.

The following tests shall constitute type tests:

(a) Visual inspection as per Clause 12.1

(b) Insulation Resistance tests as per Clause 12.2

(c) Card-level functional tests on all the cards and fail-safety tests on one card of each type.

(d) System level functional and fail-safety tests.

(e) Computerised testing for minimum two hundred thousand permutations and combinations as per Clause 12.3.

(f) Environmental/climatic tests as per Clause No. 9.0 of RDSO/SPN/144, Revision 1 (Indoor Equipment).

(g) System Diagnostics test as per Clause 12.4.

(h) System Software tests as per Clause 12.5.

11.3.2 Any other tests shall be carried out as considered necessary by RDSO.

11.3.3 Only one EI shall be tested for this purpose. The equipment shall successfully pass all the type tests for proving conformity with this specification. If the equipment fails in any of the type tests, the purchaser or his nominee at his discretion, may call for another equipment/ card(s) of the same type and subject it to all tests or to the test(s) in which failure occurred. No failure shall be permitted in the repeat test(s).

11.4   ACCEPENTANCE TEST:

11.4.1. The following shall comprise acceptance tests:

(a) Visual inspection (Clause 12.1)

(b) Insulation Resistance tests (Clause 12.2)

(c) Card level functional test on all the cards.

(d) System level functional tests.

(e) System Diagnostics test (Clause 12.4)

(f) Verification of application software vis-a-vis selection table (This shall be done by user Railway).

11.4.2. Any other tests shall be carried out as considered necessary by the purchaser.

11.5 ROUTINE TEST:

11.5.1. The following shall comprise the routine tests and shall be conducted by manufacturer on every EI and the test results will be submitted to the inspection authority before inspection. The application software in proper format shall also be submitted to the inspection authority in advance.

(g) Visual inspection (Clause 12.1)

(h) Insulation Resistance tests (Clause 12.2)

(i) Card level functional test on all the cards.

(j) System level functional test.

(k) Computerised testing for 1, 00, 000 permutations and combinations (Clause 12.3)

(l) System diagnostics test as per Cl. 12.4.

11.5.2. Any other tests shall be carried out as considered necessary by the purchaser.

12 TEST PROCEDURE

The test procedure shall be based on the system design. The methodologies to be adopted for various tests shall be decided taking into account the system design/configuration and shall be approved by the purchaser.

12.1. VISUAL INSPECTION

The equipment shall be visually inspected to ensure compliance with the requirement of Clauses 3 to 7 of this specification. The visual inspection will broadly include –

(i)   System level checking:

Constructional details
Dimensional check
General workmanship
Configuration

(ii)   Card level checking

PCB laminate thickness
General track layout
Quality of soldering and component mounting
Conformal coating
Legend printing Green masking

(iii)   Module level checking

Mechanical polarisation
General shielding arrangement of individual cards
Indications and displays
Mounting and clamping of connectors.
Proper housing of cards

12.2 INSULATION RESISTANCE TEST

This test shall be conducted between the equipment power supply line terminals and the earth. If there is a possibility of the meggering voltage reaching the cards, these will be taken out before starting the IR test.

This test shall also be carried out after the climatic tests. The measurement shall be made at a potential of not less than 500 V DC.

The IR value shall not be less than 10 Mega ohms. After the climatic tests, this value shall not be less than 10 Mega ohms.

12.3. COMPUTERIZED TESTING

The manufacturer shall provide a computer-based test set up with the required software for automatic testing.

The following tests shall be conducted with the help of this set up.

12.3.1. FUNCTIONAL TESTING

The system shall be tested functionally for all the signals with all routes, point operation, emergency point operation, route cancellation, emergency route cancellation, operation of G/F control points, level crossings and crank handle as per the selection table of the yard provided by the purchaser.

12.3.2. OPERATIONAL FAIL SAFETY TEST

These tests are conducted as per procedure given below:

(i)   After setting of points in main route & desired overlap, signal is cleared. Back locking of the route and overlap should be verified. It should also be checked that other yard functions are free.

The track circuit of the route should be dropped one by one and it should be verified that it is not possible to clear the signal. All the routes are checked one by one.

(ii)   Conditions required for route setting should be disturbed in various permutations and combinations and it should be verified that it is not possible to set the route with the disturbed conditions. Similarly, conditions required only for signal clearance (such as track circuits) should also be disturbed and it should be verified that the route is set but the signal is not cleared.

12.4. SYSTEM DIAGNOSTICS TEST

These tests shall be conducted by automatic test procedure through a PC. The diagnostic tests on the system shall be performed to test the integrity of the system software by verifying the checksum. It shall be possible to verify the application program vis-a-vis the selection table by the user, preferably through regeneration of the locking table from yard data.

The PC at the end of the test shall print out summary of the tests conducted.

12.5. SYSTEM SOFTWARE TEST

Checksum of system software and format of the application software shall be verified. In case of any change in the system software/format of application software, the same shall be validated.

12.5.1. Type test, Acceptance test and Routine test as given in para 11.3, 11.4 and 11.5 shall not be required in case of Cross-acceptance, The firm has to submit following documents to ensure that the system meets all requirements as mentioned in para 11:

(i)   Certificates of Type tests done as required by RDSO specifications.

(ii)   List of Routine tests done and sample copy of results to be submitted.

(iii)   Acceptance tests to be done at the time of inspection of equipment to be supplied.

(iv)   Performance feed back reports from user Railways.

For the verification of same, a team of RDSO officials may visit the manufacturing facilty of manufacturer in its respective Country(s). Sample tests shall be carried out, if found necessary. However, at least one set of equipment shall have to be installed in Indian Railways, to prove its performance in Indian conditions. Details of cross acceptance procedure may be referred to in concerned document of RDSO.

13.   QUALITY ASSURANCE

13.1   All materials & workmanship shall be of good quality.

13.2   Since the quality of the equipment bears a direct relationship to the manufacturing process and the environment under which it is manufactured, the manufacturer shall ensure QAP of adequate standard.

13.3   Validation and system of monitoring of QA procedure shall form a part of type approval. The necessary Plant, Machinery and Test instruments as given below shall be available with the manufacturer.

13.3.1. PLANT AND MACHINERY:

The firm should have all essential Plant & Machinery as per latest STR.

In case of CA, the above Plant and Machinery shall not be necessary to be available with Indian Partner if no TOT is taking place. In case of TOT, The same shall be required with Indian Partner of foreign firm.

In case of CA, when TOT is not taking place, the Plant & Machinery may be verified by the team of RDSO officials visiting the Firm premises.

13.3.2. All test instruments as given in Cl. 11.2.1 shall be available with the manufacturer.

13.4 Along with the prototype sample for type test, the manufacturer shall submit the Quality Assurance Manual.

14. PACKING

As per relevant clause of latest version of RDSO/SPN/144.

15. INFORMATION TO BE FURNISHED BY THE PURCHASER

(a) Approved interlocking plan, selection table and panel diagram of the station (Cl. 3.1).

(b) Whether CCIP (domino type) or VDU control terminal or both required (Cl. 5.1.4).

(c) System output required to drive field gears – relay interface or object controllers.

(d) 110V AC or DC usage for signal lamp lighting.

(e) Size of VDU monitor screen, if ordered.

APPENDIX F

ACRONYMS*

ACD	Anti Collision Device
ACE	Application Compiler Editor
ACP	Auxiliary Communication Processor
ALU	Arithmetic Logic Unit
AOCD	Absence Of Current Detector
AWS	Automatic Warning System
BDD	Binary Decision Diagram
CAST	Causal Analysis based on STAMP
CB	Counter Box
CBMC	C Bounded Model Checker
CENELEC	French Acronym for European Committee for Electrotechnical Standardization
CIU	Central Interlocking Unit
CPM	Critical Path Method
CPU	Central Processing Unit
CTC	Centralized Traffic Control
CTL	Computation Tree Logic
CTM	Central Traffic Management
DR	Dual Redundant System
EI	Electronic Interlocking
EMC	Electro Magnetic Compatibility
FMECA	Failure, Mode, Effects and Criticality Analysis
FPU	Floating Point Unit

FTA	Fault Tree Analysis
HLL	High Level Language
HOL	Higher Order Logic
IEC	International Electro technical Commission
IR	Indian Railways
IRISET	Indian Railways Institute of Signal Engineering & Telecommunictions
ISA	Independent Safety Auditor
ISO	International Standards Organization
IVV	Independent Verification and Validation
LTL	Linear Temporal Logic
MT	Maintenance Terminal
MTTR	Mean Time To Repair
NISAL	Numerically Integrated Safety Assurance Logic
NMR	N Modular System
OC	Object Controller
OFC	Optic Fibre Cable
OPCR	Output Power Control Relay
OQE	Objective Quality Evidence
PC	Personal Computer
PCB	Printed Circuit Board
PP	Panel Processor
QA	Quality Assurance
RDSO	Research Design and Standards Organization
ROBDD	Reduced Ordered Binary Decision Diagram
RRI	Route Relay Interlocking
SCM	Serial Communication Module
SDT	Safe Down Time
SE	Systems Engineering
SIL	System Integrity Level
SMV	Symbolic Model Verifier/Verification
SSI	Solid State Interlocking
STAMP	System Theoretic Accident Model and Process

STPA	System Theoretic Process Analysis
THR	Tolerable Hazard Rate
TMR	Triple Modular System
TMS	Train Management System
TPWS	Train Protection and Warning System
UIC	International Union of Railways (French Acronym)
VDU	Visual Display Unit
VHLC	Vital Harmon Logic Controller
VHM	Voltage and Health Monitoring Unit
VLM	Vital Logic Module
VPI	Vital Processor Interlocking
VSEV	Vital Serial Enable Voltage
WFM	Wayside Function Module

*Commonly used acronyms in electronics, such as ROM, have been omitted.

INDEX